Artemis I
Return to the Moon

DAVID BAKER

Front cover image: The Orion Spacecraft with its European Service Module imagined as it would look travelling to the Moon. (NASA)

Title page image: The Space Launch System rocket stands ready to launch the first Artemis mission to the Moon. (NASA)

Contents page image: The Orion spacecraft is at the core of NASA's Artemis programme returning humans to the Moon. (NASA)

Back cover image: The Orion spacecraft on Artemis I takes a picture of itself during the coast to the Moon. (NASA)

Published by Key Books
An imprint of Key Publishing Ltd
PO Box 100
Stamford
Lincs PE9 1XQ

www.keypublishing.com

The right of David Baker to be identified as the author of this book has been asserted in accordance with the Copyright, Designs and Patents Act 1988 Sections 77 and 78.

Copyright © David Baker, 2023

ISBN 978 1 80282 659 3

All rights reserved. Reproduction in whole or in part in any form whatsoever or by any means is strictly prohibited without the prior permission of the Publisher.

Typeset by SJmagic DESIGN SERVICES, India.

Contents

Introduction 4

Chapter 1 A New Dawn 5

Chapter 2 A Change of Plan 32

Chapter 3 Artemis Rising! 60

Introduction

More than 50 years after the last human boot prints were left on lunar soil, an international endeavour is on track to go back to the Moon in this decade. This is that story.

For the past 20 years, NASA has been developing the rockets and the spacecraft essential to that goal and has now set up an international team led by the United States to move towards the establishment of permanent research stations near the Moon's South Pole. It is there, in permanently shaded areas of the surface, that valuable minerals can be found and, potentially, water ice.

The goal to move humans back into deep-space operations like those conducted by the Apollo missions has been consistent, but the way in which this is to be achieved has changed dramatically over the past two decades. Multiple directions have been tried and abandoned, but the path now followed has support from politicians, the majority of the public and from the engineers and scientists working to make it happen.

The ultimate goal is to establish a working base on the Moon and to develop the technologies to make that possible. It is also about preparing a path for flights to Mars by the mid-2030s. The mission architecture for that will be very different to that for lunar base support, where personnel on the surface of the Moon are only three days' flight time away from Earth. It takes about nine months to reach Mars, and much will have to be learned before that journey is feasible.

Having established a manned space station (the International Space Station, or ISS) through an international partnership, on a facility permanently occupied since 2000, NASA is ready to lead a global coalition of national space agencies and companies to build the next two stages – a space station around the Moon, known as the Gateway, and a base on the Moon itself. The ISS will be brought down to a destructive re-entry in 2031, transferring Earth-orbit operations with astronauts to commercial companies renting research facilities in privately funded space stations.

NASA is resuming deep-space exploration with the goal, this time, of building on everything that has gone before to move humanity from the cradle of Earth to distant places in the solar system.

Chapter 1

A New Dawn

Long before the first Moon landing on 20 July 1969, NASA planned Moon bases and permanently manned space stations in Earth's orbit for the 1970s, anticipating crewed flights to Mars in the 1980s. None of that happened, principally because Congress, and the American public, saw the *Apollo* lunar landings as Cold War battles, which in their success had already won the war for loyalty from previously uncommitted nations. There was, they said, no need to continue pushing the boundaries – the future of the space programme would lie in achieving practical benefits. Big expeditions beyond Earth were simply too expensive to justify.

During the *Apollo* lunar landings of 1969–72, NASA planned to develop a reusable Space Shuttle, which was approved by President Nixon in January 1972 with two Moon missions still to go. Congress could live with that. The Shuttle would, said its advocates, lower the cost of space transportation by returning for reuse what was, in effect, a winged cargo truck, lifting satellites and carrying modules for a space station assembled in orbit and thus broadening the application of space to everyday life.

It didn't work out that way. Far more complex to operate than anticipated and expensive to fly, that 'magnificent flying machine' that was the Space Shuttle nevertheless provided opportunities for lifting

Between July 1969 and December 1972, NASA conducted six Apollo Moon landings, aided on the last three flights with Lunar Roving Vehicles transporting the astronauts far from the Lunar Module. (NASA)

Above: From 1963, NASA examined extended Moon missions through the Apollo Applications Programme, a dual manned/cargo mission represented by a contemporary illustration from Lunar Module builder Grumman. (Grumman)

Left: In this depiction, an unmanned Lunar Module delivers supplies and a truck. (Grumman)

big modules and structures into orbit to assemble the International Space Station (ISS), a cooperative venture in partnership with Europe, Japan, the Soviet Union and Canada. The value of the Shuttle lay in what it had already accomplished, and after the loss of *Columbia* on 1 February 2003, a thorough analysis of the programme laid open two possibilities: either to fund a major series of upgrades and improvements, or retire the Shuttle and pick up with deep-space operations for revisiting the Moon and going on to Mars.

The search for a broader and more enduring legacy of *Apollo* originated in depth right after the loss of the Shuttle *Challenger* on 28 January 1986. Within a year, and after the *Challenger* review commission had done most of its work, NASA established its Office of Exploration to find a way to reinvigorate the US space programme. That followed the setting up by President Ronald Reagan of a National Commission on Space, tasked with finding a bold agenda for the nation.

Studies supporting this directive followed in 1988, 1989, 1991 and 1993. However, budget pressures and the work to get the International Space Station up and running prevented action. The end of the Cold War and the absorption into the ISS of Russian (ex-Soviet) participation held things back and did little to accelerate the initiative. The prevailing attitude in Congress was a reluctance to engage in a further drawdown of public money when uncompleted programmes were pending. But all the while, both *Apollo* and the Shuttle provided possibilities for achieving great things while minimising the price. Yet, here too, there was disagreement over direction. After *Challenger* and then *Columbia* in 2003, there were deep-seated concerns about the way forward.

Exhaustive analysis of Shuttle missions revealed high risk with little possibility of improving safety at reasonable cost. A searching debate opted for retiring the Shuttle in 2010 (it actually flew until mid-2011) and replacing it with a completely new programme named *Constellation*. NASA wanted to capitalise on the invested worth by developing large rockets, launch vehicles using elements familiar on the Shuttle. Unlike *Apollo* and the Shuttle, the *Constellation* programme would inherit legacy technology to save costs and improve reliability by using proven systems.

Under the *Constellation* programme, a launch vehicle designated *Ares I* would be developed for servicing the ISS and for carrying cargo and astronauts to and from the 400-tonne facility in Earth's orbit. A considerably more powerful rocket, *Ares V*, would carry the Crew Exploration Vehicle (CEV), much like the *Apollo* capsule of old, and the *Altair* Moon lander for putting people back on the lunar surface. It would also serve to support deep-space objectives such as visiting asteroids or supporting manned Mars missions. The Roman numerals for *Ares* were allocated in commemoration of *Saturn I* and the *Saturn V*, which had been pivotal to the success of the Apollo programme.

Much of the technology for these rockets came from the Shuttle, the *Ares I* launcher consisting of a single Solid

Instead of Moon bases, after Apollo, NASA got the Shuttle, restricted to low-Earth orbit missions delivering satellites and planetary missions to space and providing work room for scientific tasks. (NASA)

The greatest achievement of the Shuttle programme was to put heavy loads into orbit, where they were assembled into the International Space Station, permanently manned since 2000. (NASA)

Rocket Booster (SRB), two of which helped propel the Shuttle on its way into orbit. A second stage was powered by a single J-2X cryogenic engine. *Ares I* would have a payload capability of 25,400kg (56,000lb) to low Earth orbit. This was approximately equivalent to the payload carried by the Shuttle Orbiter in its cargo bay. *Ares I* would only be used to send the CEV into low orbit and service the ISS. The J-2X engine was a development of the J-2 used in the second and third stages of the *Saturn V* and the S-IVB stage of the *Saturn IB*.

Initially, rocket engineers designed *Ares I* with a four-segment booster, the same as those used in pairs to launch the Shuttle. SRBs came in stacked segments, or barrels, placed on top of each other. Weight analyses showed that it would require a five-segment booster to lift the CEV, which still had to be reduced in size to bring it within the lifting capacity of that extended and more powerful booster.

At launch, the SRB powering *Ares* would ignite with a thrust of 15,000kN (3,400,000lb), firing for two minutes and thirty seconds before separating. The upper stage would fire, delivering a thrust of 1,308kN (294,000lb) for just over 13 minutes to push the payload to orbital velocity, at which point the CEV would separate. The upper stage was itself a legacy from the *Apollo* era, incorporating a common bulkhead between the separate liquid oxygen and liquid hydrogen propellant tanks. This saved space and weight and added more propellant, which raised the lift capability. Over time, *Ares I* would go through a series of different optional configurations.

The big lift

Ares I was designed to carry crew but *Ares V* was defined as a cargo-lifter, thus allowing optimisation of each rocket around clearly defined requirements and mission types. As conceived, it was to be capable of a wide range of applications, including the launch of giant astronomical telescopes, construction materials for building stations around the Moon and bases on the lunar surface. *Ares V* was never

After the loss of Shuttle *Columbia* in February 2003, NASA made the decision to close the programme, but not until after examining ways in which it could be made better, with longer Solid Rocket Boosters (SRB) incorporating an additional, fifth segment. (NASA)

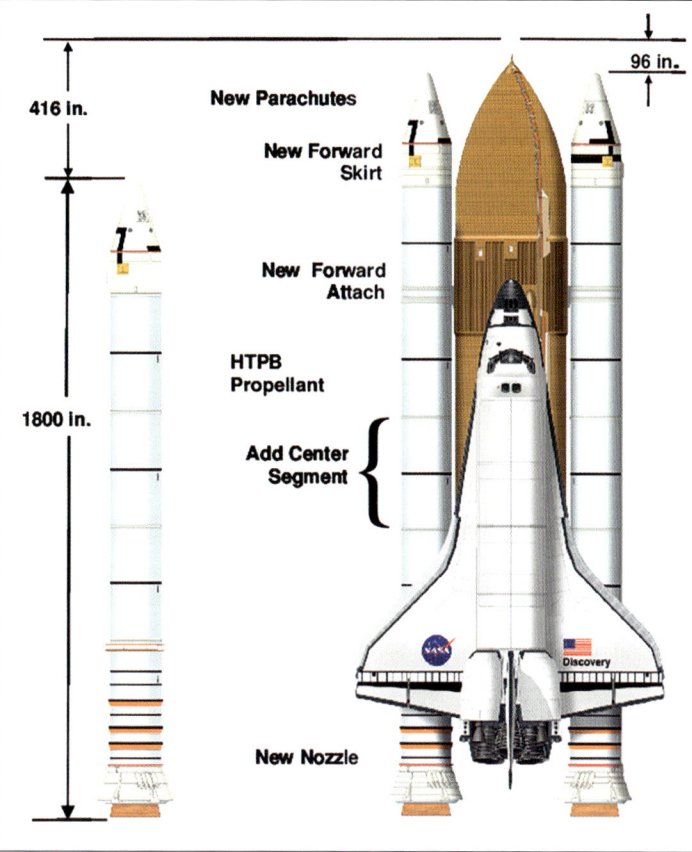

built, but it was essentially an application of legacy hardware from the Shuttle programme. This included five or six Space Shuttle Main Engines (SSMEs), fed by cryogenic propellants housed in a giant core stage containing liquid oxygen and liquid hydrogen. Two five-segment SRBs would provide boost for the first two minutes of ascent.

Whereas the *Saturn V* for Apollo could lift both the manned spacecraft and the Lunar Module for a landing on the lunar surface, the much larger vehicles for *Constellation* required the *Ares V* to lift the lander and the Earth-Departure Stage (EDS) to low Earth orbit. *Ares I* would carry the crew in the CEV to dock with the combined lander and EDS, leaving the latter to boost the assembled configuration to the Moon. The rest of the flight would be similar to *Apollo*, the whole assembly going into lunar orbit with the four astronauts going down to the surface in the lander. After a seven-day stay, they would lift off, dock with the CEV and return home.

In a mission capability very different to *Apollo*, an unmanned cargo-lifter *Ares V* could put down 20,000kg (44,100lb) on the lunar surface to support expanded base operations and construction of a suitable habitat by a crew launched by *Ares 1/Ares V* in a standard Moon-landing operation. NASA wanted to transition from an Earth-orbit capability with the Shuttle and the ISS, to deep-space operations situated at and around the Moon. This was the driving imperative behind the decision to develop two separate launch vehicles, each optimised for a specific set of roles. *Apollo* had been planned as a single-shot goal for putting boot prints on the Moon; *Constellation* would aim to establish a more sustainable presence for scientific research.

The philosophy was to have a single set of hardware also applicable to Mars missions. For that, it was desirable to allow a wide margin between exposure to harmful radiation on the way to Mars and on the way back, minimising the flight time to six months each way. The crew would remain at Mars for 18 months, insulated by a protective structure. This would help to screen the crew from solar radiation, to which they would be exposed on any space flight outside the protective envelope of the Earth's magnetosphere. The Earth-orbiting ISS is well within the magnetosphere and astronauts in the station are thus protected from deep-space radiation.

Ares V would do the heavy-lift, but it had a potential competitor in a monster rocket called *Magnum*, proposed by Boeing. With two liquid propellant side boosters instead of solid propellant SRBs, each *Magnum* booster would also have jet engines and small wings to enable it to fly back for a conventional landing. A very wide range of concepts offered by industry and NASA field centres was proposed, but the ambitious needs of their goals rendered most of them incapable of supporting the objectives for *Constellation*, which in reality still had to be defined.

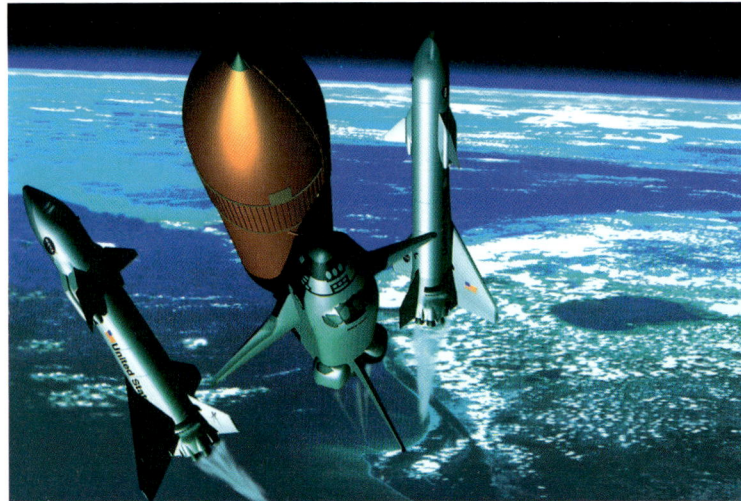

Left: To improve efficiency and lower costs, NASA looked at winged, fly-back boosters, which could land on a runway after use. (NASA)

Below: Turning to deep-space goals after the Shuttle, critical elements of that reusable vehicle, including the Space Shuttle Main Engine (SSME), were to form the primary equipment for a new transportation system. (NASA)

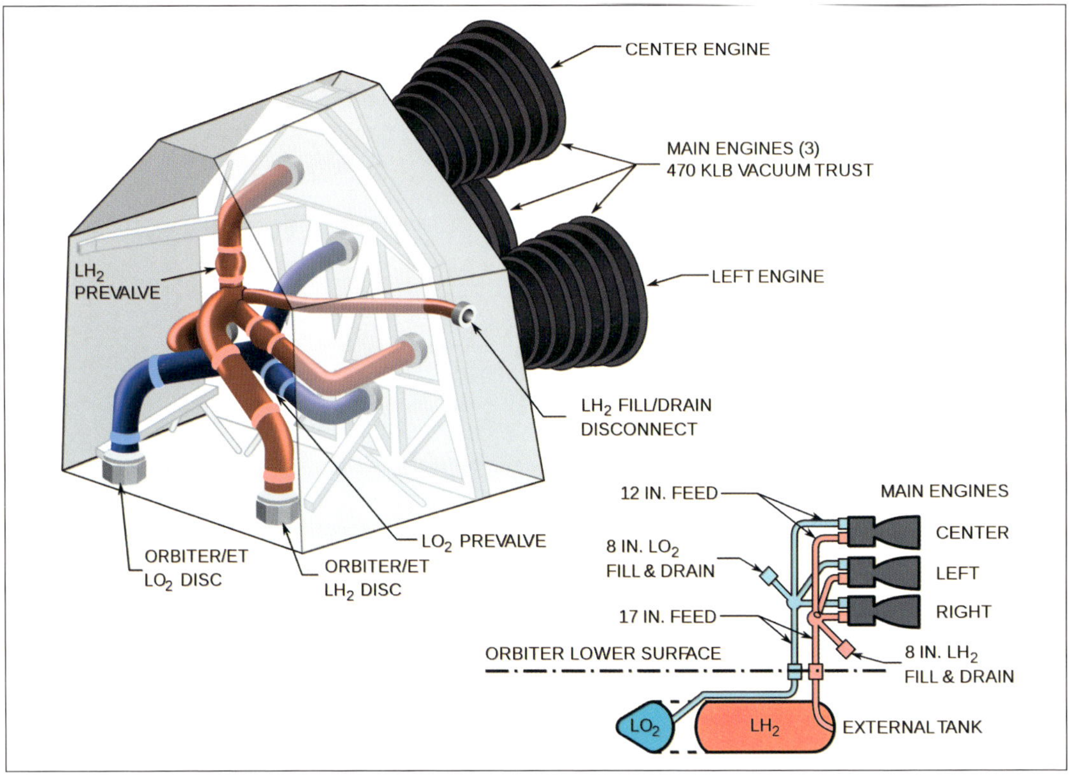

The Shuttle RS-15 main engines had been designed for re-use after the Orbiter returned to Earth, but these would be repurposed into the main propulsion for a new generation of heavy-lift rockets. (NASA)

Eventually, NASA would use several technologies from the Shuttle for its new deep-space projects, including the Orbital Manoeuvring System (OMS) engines seen here with red covers over their exhaust nozzles. (NASA)

The overall expectation with *Constellation* was that the Shuttle would be retired in 2010 and that US participation in the ISS would end in 2016, with *Ares I* and the CEV providing support to the ISS in the intervening period. However, a lot of development work was required to produce the *Ares* launchers, the CEV and also the lunar lander itself, which would soon become known as *Altair*.

With a plan to put astronauts back on the Moon in 2018, design of the *Altair* lander moved ahead quickly. With a total weight of 45,864kg (101,113lb), about three times that of the *Apollo* Lunar Module, it was required to support four men on the surface for up to seven days. Like the Lunar Module, it would consist of two separate sections, the lower section carrying the descent engine and four landing legs with foot pans, and the upper section carrying the crew compartment and another engine for carrying the astronauts back to the CEV, with the mother-ship left in lunar orbit.

The key element in this architecture was the CEV, carrying a crew of four. Pivotal to missions envisaged was its ability to support crew and cargo movements to the ISS between 2011 and 2016 and then to transition to deep-space operations. ISS support operations would employ a Block 1 configuration, with Moon missions using a Block 2 configuration and Mars flights getting a Block 3. This, however, was an oversimplification of the requirements for three very different types of mission and even the shape and size of the crew vehicle had been subject to an evolutionary process.

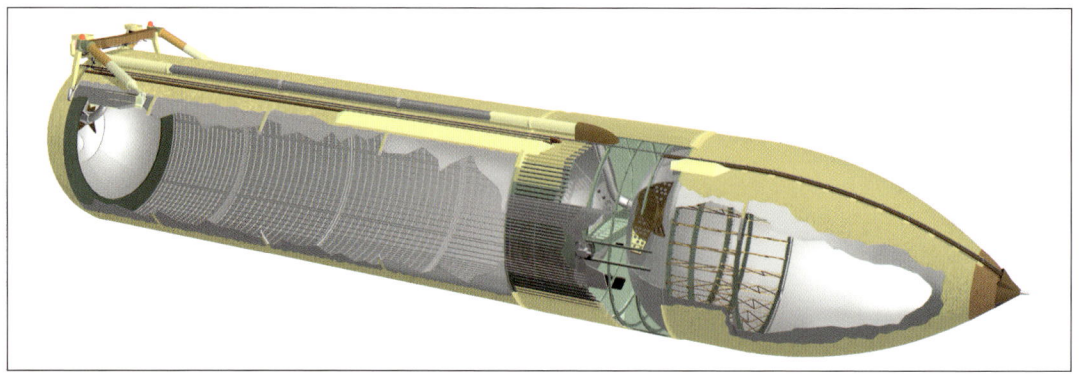

Above: The technology that produced the Shuttle's External Tank would apply its cryogenic, hydrogen/oxygen propulsion system to a new high-energy rocket stage. (NASA)

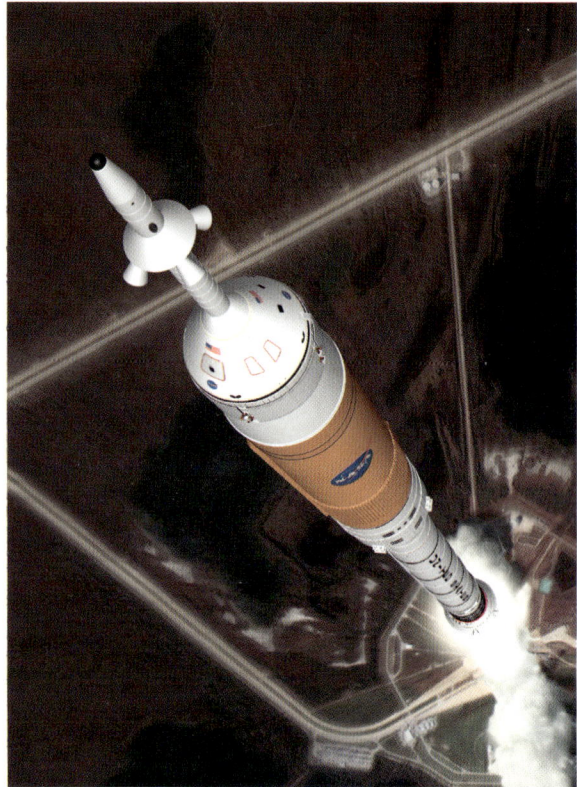

Left: NASA leaned heavily on Shuttle elements for two new launchers; the *Ares I* comprised a single Solid Rocket Booster with a cryogenic upper stage for placing a manned spacecraft in Earth orbit. (NASA)

The design of the crew vehicle grew out of the *Apollo* Command Module (CM), the optimum shape for returning a manned spacecraft to Earth. The *Apollo* CM was 3.9m (12.75ft) in diameter, with a sidewall slope angle of 32.5 degrees and an internal pressurised volume of 10.4m^3 (367ft^3) for three astronauts. NASA studied optional layouts and various sizes based on the conical shape. Initially, it had been considered that the CEV would be attached directly to a lander and support a crew of six. When the combined *Ares I*/*Ares V* flight mode had been selected and Altair was chosen as a lander, the CEV baseline was for a crew of four.

With a diameter of 5.5m (18ft) and a pressurised volume of 30.6m^3 (1,080ft^3), the four-place CEV would provide more than twice the space for each individual as the three-place *Apollo* spacecraft. Similar to the *Apollo* spacecraft, to support the CEV a service module would be attached with solar cell arrays deployed for providing electrical power. The fuel cells providing power for *Apollo* were possible because of the operating lifetime of the spacecraft, designed for lunar missions of up to two weeks.

In bringing reactant oxygen and hydrogen together over a catalyst to provide electrical power and water as a by-product, fuel cells would have been most efficient. However, because the *Constellation* CEV was also designed to supply the ISS and remain at the station for up to nine months, it was impractical to carry cryogenic reactants for that length of time. The service module would also provide propulsion for orbit and attitude changes and provision for communications.

Gathering momentum

Constellation was a hybrid programme exploiting research, development and operating experience with both *Apollo* and Shuttle systems. The idea was timely; only the political will to make it happen remained and that was swift in coming. On 14 January 2004, President George W. Bush made a speech announcing that America was going back to the Moon, and would be building the rockets and the spacecraft to carry humans to other destinations in deep space, including Mars.

On 2 May 2005, NASA Administrator Sean O'Keefe inaugurated the Exploration Systems Architecture Study, or ESAS. It was to conduct a complete assessment of the CEV and its development plan from the ISS supply ship to support Moon missions. It would list requirements for crew and cargo launch systems for the Moon and Mars, put together a lunar exploration architecture and identify key enabling technologies. By the end of the year, the ESAS team had a definitive programme plan.

Much of the preceding description of the CEV and *Ares I* was adopted and refined, but a definitive design preference for the *Ares V* cargo-lifter emerged. The preferred option was for a launch vehicle consisting of a core stage, as defined previously, with the same diameter as the external tank for the Shuttle – 8.38m (27.5ft) – but much longer and with 38 per cent more propellant. Powered by five RS-25 engines derived from the Shuttle, at lift-off the core stage would produce a thrust of 9,296kN (2.09 million lb), to which would be added the 33,360kN (7.5 million lb) of thrust from the two five-segment solid rocket boosters.

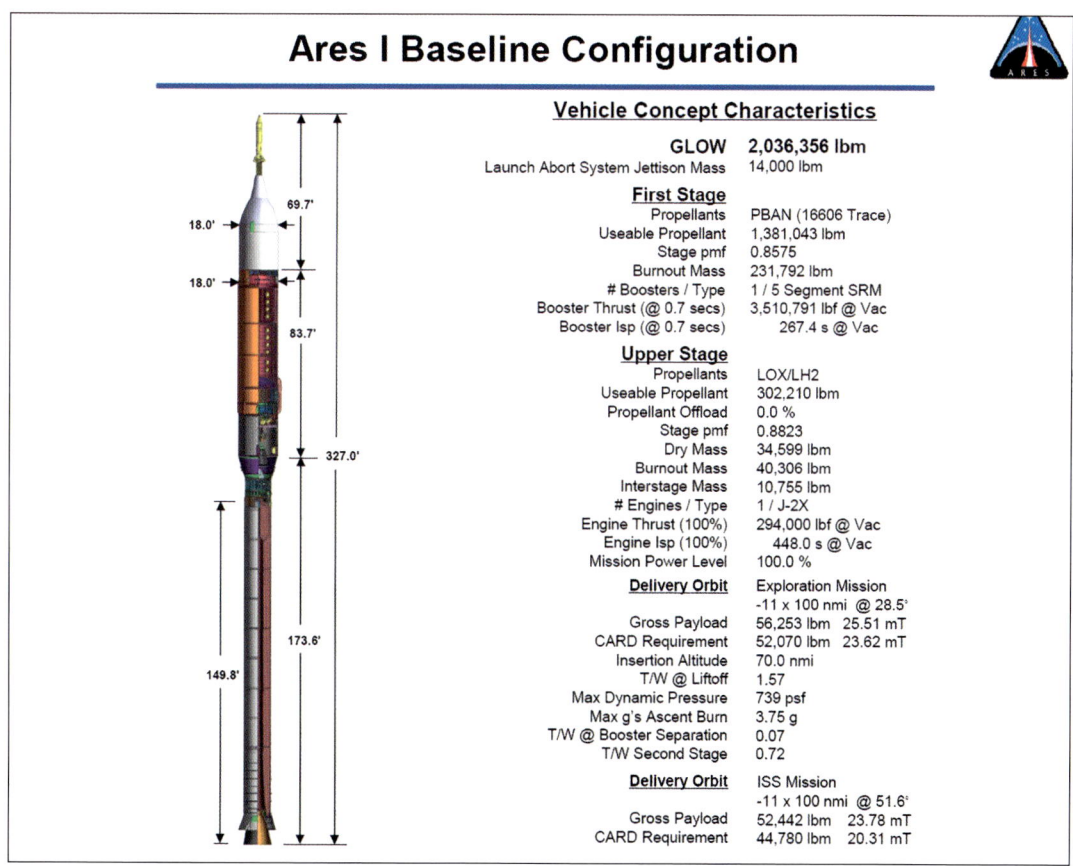

The *Ares I* was designed to be a low-cost launcher for servicing the International Space Station and carrying astronauts back and forth to the orbiting laboratory. (NASA)

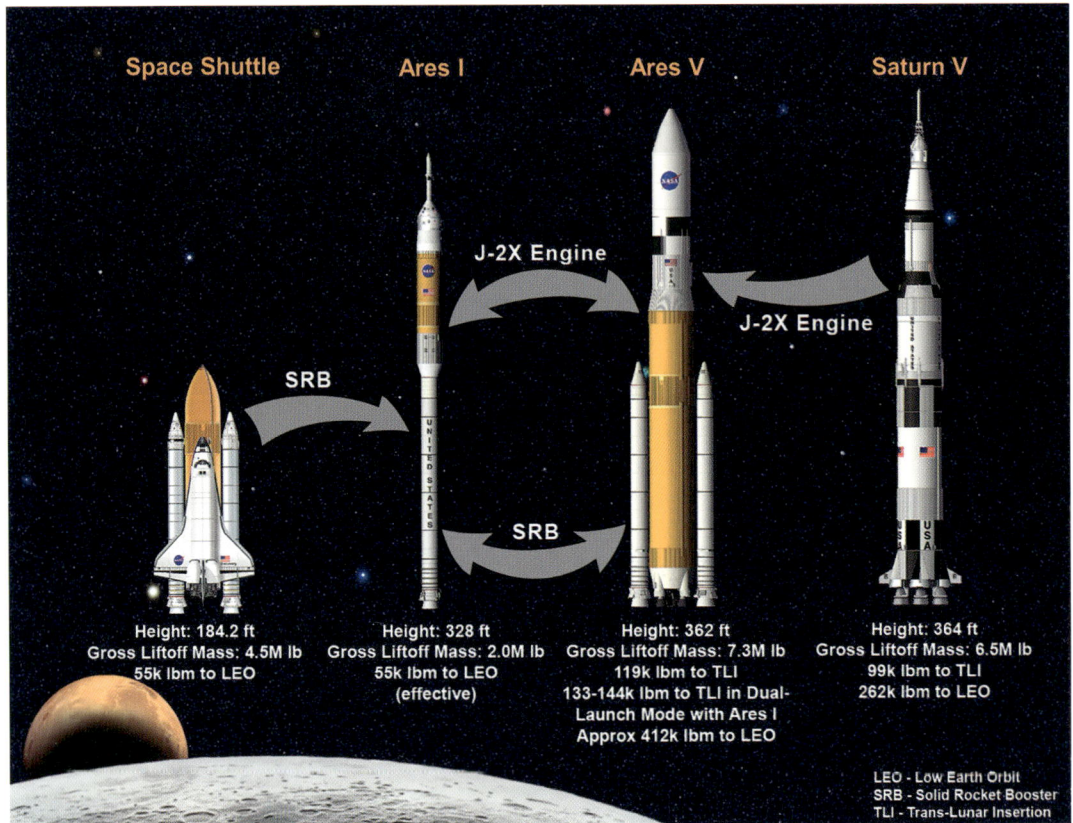

From the Shuttle, NASA scavenged main rocket motors, solid boosters and electronic control systems and avionics to produce the *Ares I* and the *Ares V* heavy-lifter, here depicted with the Apollo-era *Saturn V* for comparison. (NASA)

The ascent phase would begin with ignition of the five RSA-25s at the base of the core stage, followed several seconds later by ignition of the solid rocket boosters and lift-off. The boosters would separate at 2 minutes 12.5 seconds and the core stage would continue to burn for a total of 6 minutes 48 seconds, placing the Earth Departure Stage (EDS) and payload into a suborbital trajectory. On separating from the upper elements, the core stage would fall back into the atmosphere and burn up over the Pacific Ocean.

Powered by two uprated J-2S engines, the upper stage (EDS) would burn for 3 minutes 38 seconds, placing itself and the attached payload into a shallow, elliptical path around the Earth. This would be circularised by a second burn. To depart Earth orbit and head to the Moon, the EDS would ignite for a third time, in a burn lasting 2 minutes 34 seconds. This would occur after a CEV launched by *Ares I* had been sent to rendezvous with the *Altair* lander it was carrying, or go direct to the Moon if it was lifting unmanned cargo with a system for getting down to the surface.

By early 2006, work was under way on a wide range of technologies and a broad blueprint had been developed as a roadmap. Under this plan, humans would re-establish a presence on the Moon with the first in a series of short-duration sortie missions. This would be quickly followed by the development of a lunar outpost, using ground vehicles for speedily traversing the surface, thus establishing a permanent base for scientific research. All of this would be pursuant to an ultimate goal of repeating this cycle on Mars.

The ESAS team put together a set of Design Reference Missions (DRMs) that would guide planning for the necessary technology and group flight objectives into pre-defined packages. The primary objective of DRM-1 was to send a crew of three to the ISS for a six-month stay and then return them to Earth. DRM-2 would provide unpressurised cargo transfer to the ISS using a similar trajectory to DRM-1, while DRM-3 would do the same with pressurised cargo.

Lunar operations would follow the DRM-4 flight plan described earlier and include initial sorties to the surface, much like the last *Apollo* flights. The architecture would support four people on the Moon for seven days, with activity outside in pressure suits on each day. To establish a preliminary outpost, cargo would be required, and for that, DRM-5 would profile an *Ares V* delivery package bringing power supplies, advanced communications equipment and a simple mobility system.

The establishment of a permanently occupied base was the template for DRM-6, in which a combined crew and cargo delivery would be made to a site near the South Pole. Scientists had already identified several areas at that location where water ice was likely to be found in permanently shaded places and where other valuable resource materials could be extracted. The DRM-6 profile envisaged a crew of six remaining at such an outpost for six months, replaced after that by a new expedition of six crew members. The overall mission would be identical to a sortie flight but for the extended stay between landing and leaving.

Implicit in the strategy was the potential to pre-position a lander at the surface as a standby in the event the primary vehicle failed to get off the Moon. The plan did include a pre-positioned lander ready for a crew that would receive cargo instead of an ascent section for lifting off, rendezvousing with the CEV and coming home. The options were expanded by the multiple use of launchers and payloads, and these were admirably suited to the high demands placed on lunar operations, from where it takes three days to get home. Much more challenging were the requirements for expanding the expeditionary profile to a Mars mission.

DRM-7 defined the requirements and the flight plan for such an endeavour, beginning with a full pre-positioning of support elements 26 months before the crew flight. This was the interval between launch windows to Mars and ensured that resources were available for a six-person crew to make this 2.5-year mission, consisting of transit flights of six months and surface time of 18 months. The crew would begin by launching into low Earth orbit in a CEV and move into a Mars Transfer Vehicle (MTV). The CEV would be shut down into a quiescent mode and only periodically checked out as it was carried to and from an orbit around Mars.

Ares V was viewed as the heavy cargo-lifter and was envisaged for carrying up the deep-space habitats and lunar lander ready for occupation by astronauts lifted on the *Ares I*. (NASA)

Surface operations were not specified by the ESAS *Constellation* group, nor was the kind of Mars lander required, although there had earlier been a preference for using the *Altair* Moon lander for that role. The crew would set down on the surface at the spot where support equipment had been deposited two years earlier by the single flight of an *Ares V*. The MTV would not fly within Mars's atmosphere but remain in orbit until the crew returned to dock with it and begin the flight back home. A day before reaching Earth, the crew would move into the CEV, undock and align their flight path for re-entry, leaving the MTV to fly past.

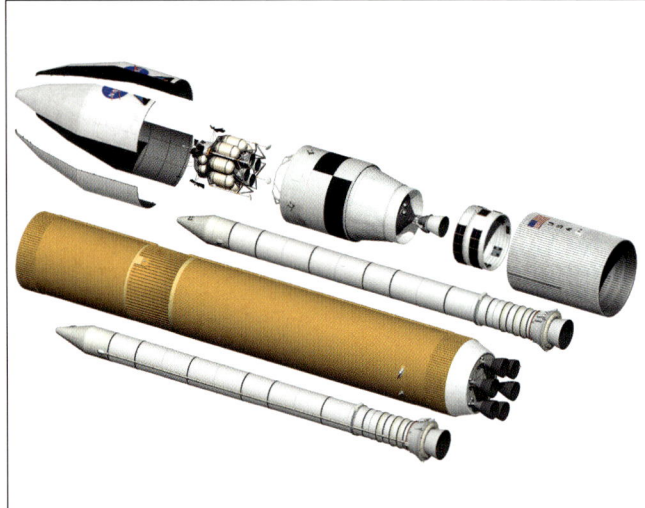

Left: **Ares V maximised availability of existing technology from the Shuttle programme, including the cryogenic engines, the Solid Rocket Boosters and an effective upper stage, here seen with the Altair lander. (NASA)**

Below left: **Careful examination of this chart shows that the amount of energy in km/sec expended getting into low-Earth orbit is far greater than the energy required to reach the Moon and Mars from that location. Cislunar (Earth-Moon) transport was key to moving around the solar system. (NASA)**

Below right: **NASA examined several options for heavy-lift into Earth orbit, including a Shuttle where the Orbiter is replaced by a cargo pod. (Boeing)**

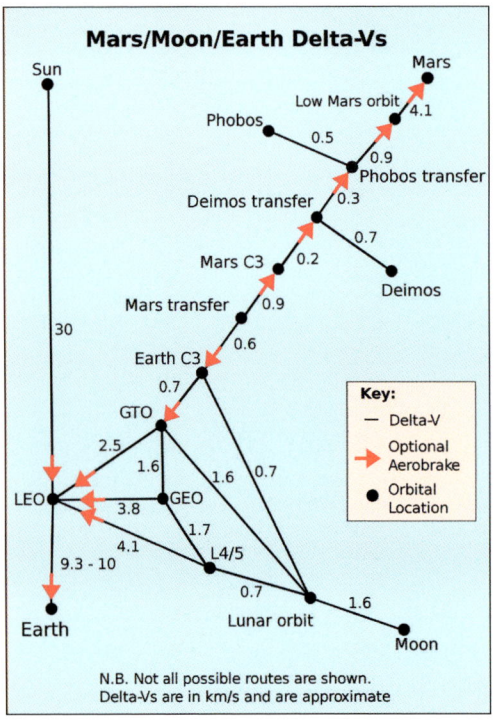

Above: The technology required to build a very large core stage for *Ares V* came directly from the Shuttle External Tank. (General Dynamics)

Right: As an alternative to *Ares V*, the *Magnum* concept was proposed by Boeing but fell victim to cheaper application of existing technology. (Boeing)

The bigger picture

Underpinning these aspirations was an obscure hope of offloading responsibility for crew and cargo delivery to commercial companies. It was just a low-level possibility, but it increasingly appeared to make sense. Right at the beginning of the *Constellation* programme, the entrepreneur and disruptor Elon Musk was battling hard to get a private launch company, SpaceX, up and running. Musk wanted to compete with governments and big corporations to offer low-cost transportation to space. To do this, he attracted a few very bright engineers and rocket scientists from the big aerospace companies, promising a fresher, quicker and leaner operation to fast-track ideas and new possibilities in a radical approach to space flight.

From this core pool of talent, Musk established a nascent research and test programme, which sought to reset the way rocket development and space programmes could evolve, away from overarching government constraints and free from corporate bureaucracy. After several failures, SpaceX turned a corner on the last attempt and achieved a success. But that would not be until 2008, although in the early years of the *Constellation* programme much consideration was given to contracting out such important services, which, until the end of that decade, was only an aspirational possibility.

Preparations for commercial activity began with the Commercial Orbital Transportation Service (COTS) programme, for which NASA solicited proposals from industry in 2005. SpaceX submitted its *Dragon* spacecraft design and on 18 August 2006 it received a contract to develop such services. A second contract was awarded to Kistler Aerospace but, when that company failed to fulfil its obligations, the agreement was transferred to Orbital Sciences Corporation with its own concept.

Above: NASA wanted the Crew Exploration Vehicle (CEV) to service the International Space Station. This depiction imagines it docking at the facility. (NASA)

Left: The space station was key to preparing humans for deep-space exploration and the CEV is shown docked to the axial port on the US module. The Japanese module is to the left and the European *Columbus* module to the right.

For its deep-space role, the CEV would await the launch of an *Altair* Moon lander by *Ares V*. (NASA)

Launched on an *Ares I*, the CEV docks with the *Altair* lander ready for a boost to the Moon. (NASA)

The upper stage of *Ares V* sends the docked configuration to the Moon. (NASA)

Rising like a phoenix from near-oblivion, SpaceX followed a string of rocket failures with its *Falcon* launcher to achieve success and an increasingly confident following in NASA. On 23 December 2008, SpaceX received a full contract worth $1.6 billion to fly at least 12,000kg (44,000lb) to the ISS across 12 cargo supply flights. Considerable sums of government money were awarded to commercial contenders to bolster funding provided by the start-ups in an expanding environment of entrepreneurial activity known as 'New Space'. The shared partnership between private and public operations would forge the future for manned as well as unmanned opportunities.

Meanwhile, on the government side of the programme, NASA had made strides in getting the CEV under way. A contractor needed to be found and, for the first time in more than 30 years, the agency was fishing for a manufacturer to build its next-generation manned spacecraft. Gone were the old bidders. Contracted in 1972, the Shuttle Orbiter had been built by North American Rockwell, re-branded as Rockwell International. Before that, from 1961 the same company had built the *Apollo* spacecraft and from 1962 Grumman had built the Lunar Module. Before them all, McDonnell Aircraft Company had built the *Mercury* spacecraft from 1959 and the *Gemini* spacecraft from 1961. As they existed in their original form, all were gone and corporate objectives had shifted.

Above: After landing on the lunar surface and returning to the CEV, the two vehicles dock, with the crew and surface samples being transferred for the voyage home. (NASA)

Left: The CEV fires out of lunar orbit to begin a three-day journey back to Earth. (NASA)

So it was that when NASA began searching for a contractor to build the CEV, it started work with a Northrop/Grumman team and with Lockheed Martin for concept definition and trade studies. On 31 August 2006, Lockheed Martin won the contract to build the CEV, which it named *Orion* under an initial agreement estimated at $3.9 billion dollars at 2006 prices. Lockheed claimed that the contract would pay for 2,300 workers, mostly employed at its Denver, Colorado, facility. While retaining the first-flight date of 2014 for the launch into Earth orbit of a test flight on an *Ares I* rocket, the first Moon landing date quickly slipped from 2018 to 2020.

As refined in the basic design outlined by NASA during its CEV studies, *Orion* would be similar to, and borrow much from, the *Apollo* Command and Service Modules. It would have a Launch Abort System (LAS) for lifting the crewed vehicle away from an ascending stack if it ran amok at any point from lift-off until well into space. *Orion* would benefit from the revolution in avionics and automated flight control systems, with the aviation world providing equipment, systems technology and operating protocols to revolutionise the way it would operate compared to *Apollo*.

As the programme evolved, manufacturing was assigned to the Michoud Assembly Facility in New Orleans. Although the electronics and operating systems for controlling and operating the spacecraft were new, *Apollo* played a long shadow across the design of the spacecraft hardware, especially the heat shield, which would face even greater demands than had the Command Module of the *Apollo* missions.

For thermal protection from the searing heat of re-entry, where temperatures could reach almost 2,760 degrees C (5,000 degrees F), NASA chose to inherit the ablative shield concept developed by Avco for the Command Module. It consisted of a phenolic epoxy resin known as Avcoat, impregnated

Right: Seen here with NASA boss Sean O'Keefe (right), President George W. Bush announced the Constellation Program on 14 January 2004. (NASA)

Below: To speed development, NASA looked at an *Ares IV* concept incorporating the upper stage from *Ares I* to provide a payload capacity of 41,000kg (90,420lb) to low Earth orbit. (NASA)

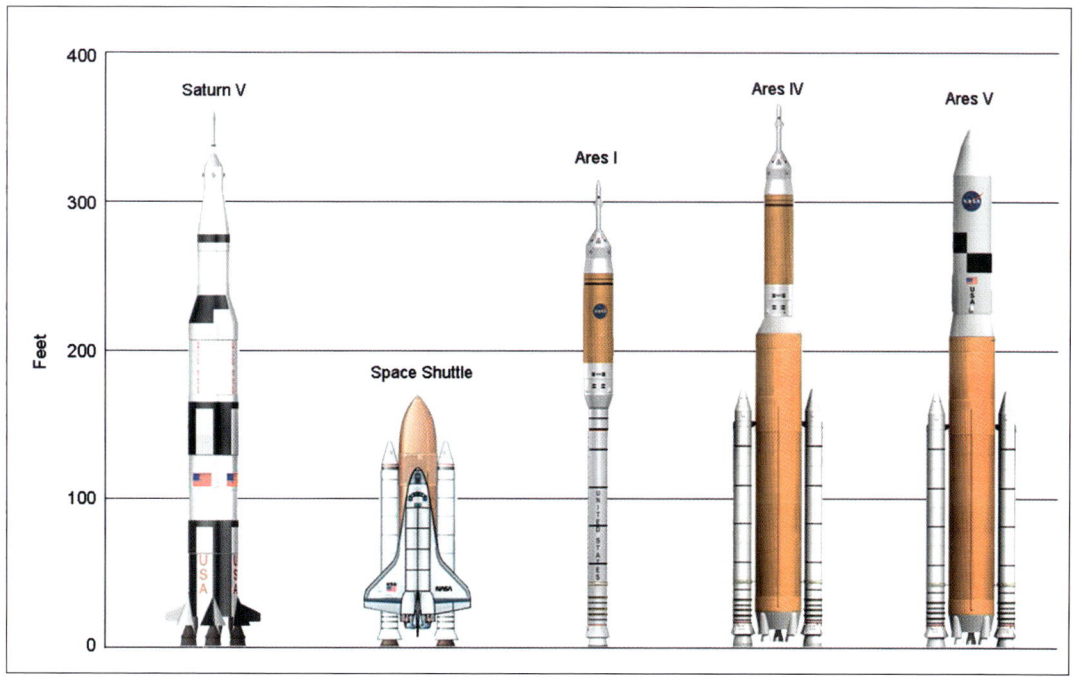

An open rover for work on the Moon, tested at Black Point Lava Flow in Arizona. (NASA)

Real-world geology excursions were tried out on three-day simulations of a field trip on the Moon. (NASA)

Above: The Surface Exploration Vehicle was designed for long-duration journeys to distant locations for scientific surveys on the Moon or Mars. (NASA)

Right: Studies leaned toward mobility, with Moon bases capable of physically relocating to areas of geologic interest. (NASA)

Automated or remotely controlled vehicles could be deployed for carrying large or massive geologic survey equipment. (NASA)

Above: An unmanned freighter arrives at a remote geologic site, called in as new equipment is requested by an on-site survey team. (NASA)

Below left: The abundance of various geologic materials in lunar soil, known as regolith. (LPRI)

Below right: A target of great interest for future lunar explorers is the large depression known as the 'Aitken Basin', about 2,500km (1,600 miles) in diameter. (LPRI)

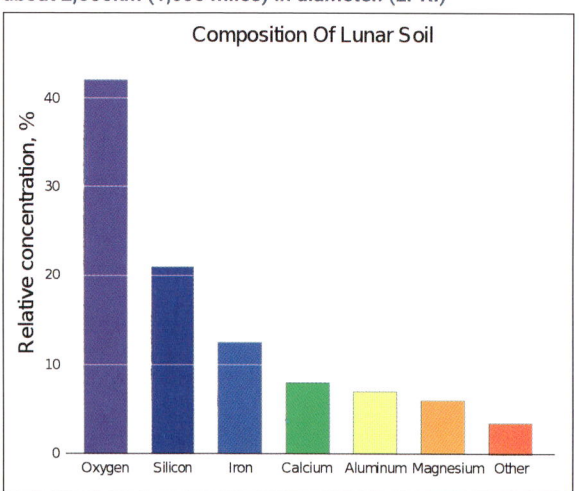

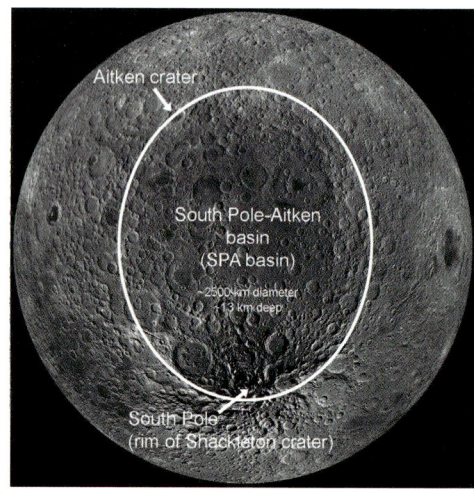

with silica fibres and poured into a fibreglass-phenolic honeycomb consisting of 320,000 individual cells, into which it set hard. Designed to char away and carry the heat with it, the ablative heat shield would take the brunt of re-entry and be the only means by which the crew could survive. There were challenges in this. The 27 per cent increase in the base diameter of the *Orion* spacecraft compared to the *Apollo* Command Module raised the surface area by 63 per cent. The upscaling of the Avcoat shield would prove troublesome when NASA came to fly the first *Orion* capsule unmanned in 2014, and was

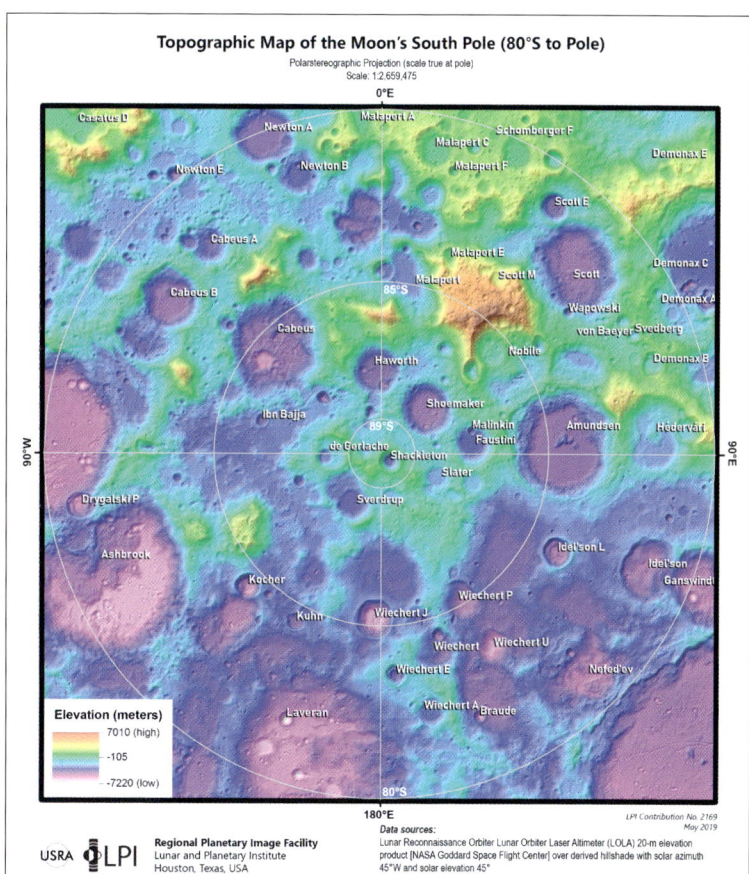

Right: A topographic map of the Moon's south polar region, where interest is focused for the next series of manned lunar expeditions. (LPRI)

Below: NASA always viewed *Constellation* as a stepping-stone to a Mars mission. The CEV is seen docked to a Mars Transfer Vehicle. (NASA)

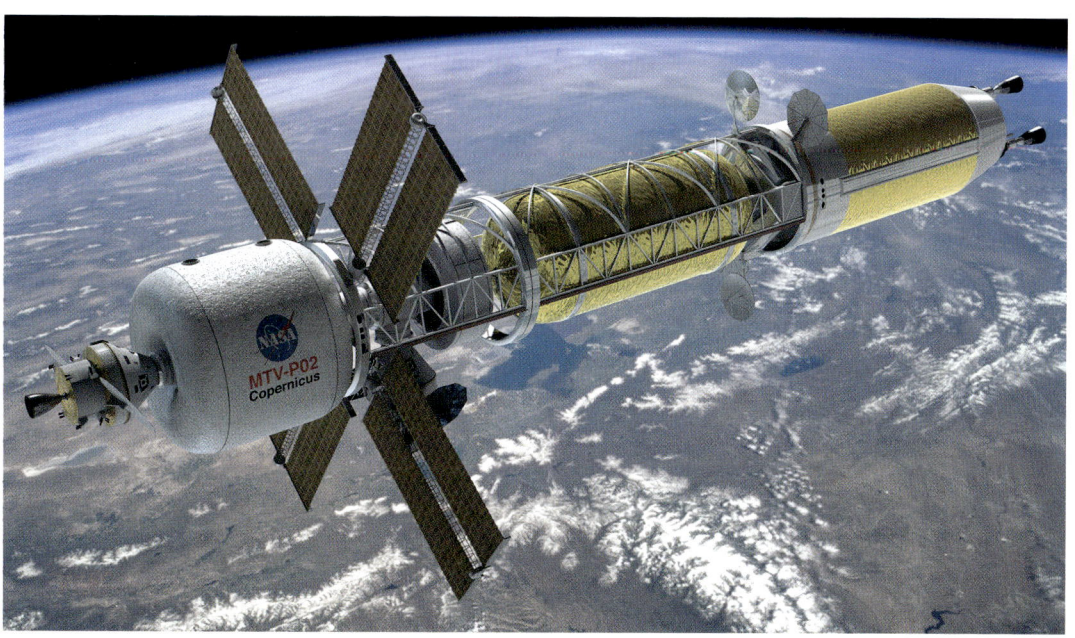

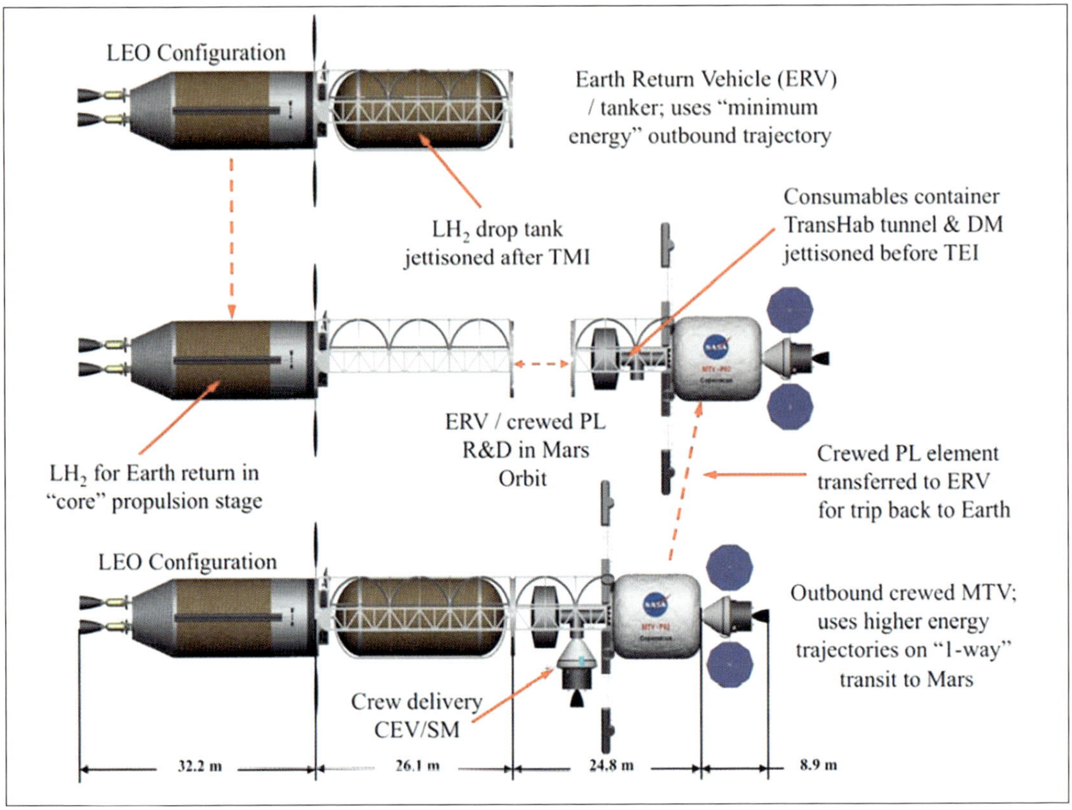

Above: The basic elements of the Mars Transfer Vehicle, with assembly in low-Earth orbit (LEO) before departing. (NASA)

Left: Founder of SpaceX, entrepreneur Elon Musk had a flurry of failures at the start of his illustrious career, and is seen here after one such attempt ended in disaster. (David Baker)

just one of several worrying concerns when NASA began testing the technology and legacy concepts. What had worked with *Apollo* would not necessarily be possible with a much larger vehicle.

Engineering complexities were not the only problems. Costs began to soar and projections of future funding requirements began to show disparities between the demanding schedule of its programme plan and the realities emerging from real-world work. The recognition dawned that NASA could simply not afford to build two new launch vehicles and two new spacecraft on the timeline projected by the White House. The *Altair* lander programme, a major development in itself, was consistently running behind schedule.

Right: With increasing success, SpaceX bid successfully for a commercial programme to deliver cargo to the International Space Station. NASA boss Charlie Bolden congratulates Elon Musk. (SpaceX)

Below: The first demonstration flight of the SpaceX Crew Dragon in May 2022 as it is docked to the International Space Station. (NASA)

Left: The Dragon capsule is recovered at sea after a successful splashdown, returning to Earth with experiments and trash from the International Space Station. (SpaceX)

Below left: A second commercial cargo provider is Orbital Sciences (now a subsidiary of Northrop Grumman) with its *Antares* rocket launched from Wallops Island, Virginia. (NASA)

Below right: A *Cygnus* freighter docks with the International Space Station. (NASA)

The *Orion* Service Module, meanwhile, was proving troublesome and costly. Discussions with Lockheed Martin exposed a gulf between the value of the contract and the money required to get the job done. In 2008, senior NASA management began to hold informal talks with the European Space Agency (ESA) about a possible venture that would see them adapt a logistics module developed for the ISS into the service module for *Orion*, saving money for the US taxpayer.

Since signing a partner agreement with the station in the mid-1980s, the European Union had committed to building modules and to supplying elements for the ISS in exchange for European astronauts working in space. One of those elements was the Automated Transfer Vehicle (ATV), an

Astronaut Candy Coleman in Mission Control, Houston, acts as capsule communicator with the International Space Station as *Cygnus* is moved from one port to another. (NASA)

An enhanced and upgraded *Cygnus* freighter, capable of carrying three tonnes (6,615lb) of cargo, approaches the International Space Station. (NASA)

Mission Control at NASA's Johnson Space Center during Dragon rendezvous operations at the International Space Station. (NASA)

unmanned cargo carrier launched by Europe's own *Ariane* launch vehicle to automatically dock with the ISS.

The ATV had all the technology elements for ESA to apply that capability to *Orion*, providing a readily available service module. Admittedly, that would take work away from Lockheed Martin, but would enjoy with *Orion* the same reciprocal arrangement that had worked so productively with the ISS. In 2008 and early 2009 it was just an idea. But as dark clouds gathered over the future of *Constellation*, over time it would save the programme.

Left: Lockheed Martin won the contract to build the Crew Exploration Vehicle (*Orion*) as seen here with its circular, foldable solar arrays of the design first used on its Mars landers. (NASA)

Below left: The heat shield selected for the CEV was derived from Avcoat, used on the *Apollo* Command Module and seen here during manufacture in the 1960s. (Avco Corporation)

Below right: Technicians inject an epoxy resin into individual honeycomb cells on the heat shield for the *Apollo* Command Module, an ablative compound adopted for the CEV (*Orion*). (Avco Corporation).

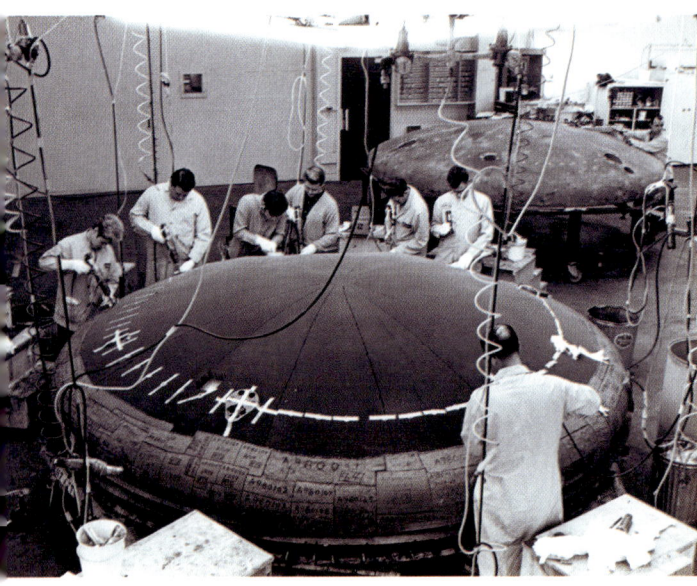

Above: Seen here in the left foreground with its rocket motors firing, Europe's ATV was suitable for providing vital power, communications, environmental control and propulsion for NASA's CEV (*Orion*). (ESA)

Right: The European Space Agency provided the Automated Transfer Vehicle (ATV) for the International Space Station and conducted discussions with NASA about using it as the Service Module for the CEV. (ESA)

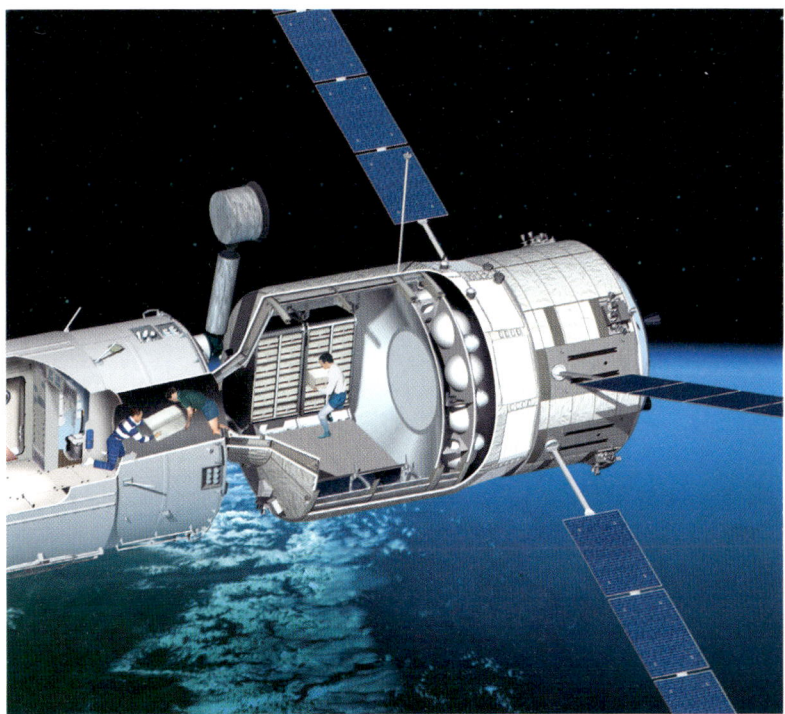

Chapter 2
A Change of Plan

On 20 January 2009, Barack Obama, the Democratic Party nominee and winner of the 2008 national election for the highest office, was inaugurated as the 44th President of the United States. Suspicious of the overarching plan laid out by his predecessor and political opponent, and wanting to conduct a full review of human space flight plans, Obama set up a commission under Norman Augustine, former CEO of Lockheed Martin and one of the most respected leaders in the aerospace industry.

Many in and around NASA had been aware of the financial pressures surrounding an agency given high targets but low budgets. There was simply not enough money to pay for the *Constellation* programme proposed by President George W. Bush and authorised by Congress. In the preceding five years, a great deal of technology had been developed, but numerous critical elements were several years behind the timeline to have the CEV carrying humans by 2014 and Moon landings from 2018.

While the Augustine Commission set about its review, Obama appointed former astronaut Charles F. Bolden Jr as NASA Administrator, the top job at the space agency. His deputy was to be Lori Garver, with both appointments effective from 17 July 2009. It is the Administrator's job to carry out the policies set down by the White House and as a government office, NASA can only carry out what it is allowed to by the government. However, it is the legislature – the Senate and the House of Representatives – that has power to pass laws, approve programmes and provide the finances.

Sensing growing opposition to the increasing cost predictions over *Constellation*, and aware that the Augustine Commission was raising the spectre of yet another redirection in the national programme, Lockheed Martin conducted internal studies about alternative mission objectives for *Orion*. By mid-2009, it had formulated a plan to save *Orion*, should NASA switch from the lunar landing goal to a focus on exploring the asteroids. To do so would eliminate the costly *Altair* lander and base a future

Key figures would topple NASA's manned space flight plans. Faced with mounting costs and technical delays, President Barack Obama commissioned a report into the *Constellation* programme. (NASA)

Above left: President Barack Obama (left) receives a flight jacket from NASA's Director of Flight Crew Operations, Janet Kavandi, and meets the final Shuttle crew of (background left to right) Doug Hurley, Sandy Magnus, Chris Ferguson and Rex Walheim. (White House)

Above right: Aerospace magnate, Secretary of the Army (1975–77) and CEO and chairman of Lockheed Martin, Norman Augustine issued a critical report on the manned space flight plans. (NASA)

Right: Second in command at NASA from 2009 to 2013, Lori Garver pushed hard for relying on commercial companies and entrepreneurs to form the backbone of NASA'S manned space programmes. (NASA)

programme on *Orion*, by which two spacecraft docked nose-to-nose could rendezvous with an asteroid for detailed analysis and sample retrieval.

The combined, low-level discourse between Lockheed Martin and NASA sought to move ahead of any damning conclusions about the cost overrun and schedule slip with *Constellation*. There were already serious development problems with the two *Ares* launch vehicles, but they were on a critical path to any operational use of *Orion*. Their development was unavoidable, given that there was no other launcher to lift *Orion* into space.

Asteroids, and the potential threat they might pose to life on Earth should a large object strike the planet, was a topical conversation among scientists and a significant proportion of the general public. Films about the extinction of the dinosaurs 65 million years ago, believed to have been caused by an asteroid striking Earth, garnered public awareness. Serious efforts were funded by several governments around the world to use space technology to track potential Earth-intersecting objects and to know more about their composition and structure, perhaps even in the interest of developing techniques for deflecting them away from a collision with Earth. Because of these factors, the investigation of asteroids had greater traction with the public than a return to the Moon.

Above: President Obama brings the First Family to see the last Shuttle *Atlantis* at the Kennedy Space Center before its last flight in July 2011 on STS-135. (NASA)

Left: An outstanding engineer and manager, William Gerstenmaier joined NASA in 1977 and is here seen (foreground) with a wind tunnel model of the Shuttle in 1978, three years before its first flight. (NASA)

Over the preceding two years, a wide range of problems had arisen with the *Ares* launchers, and concerns about a range of fundamental issues appeared insoluble. Taking a single Shuttle-type SRB and turning it into what cynics described as a 'single-stick launcher' (*Ares I*) introduced high vibration levels and acoustic spectra incompatible with the manned *Orion* spacecraft on top of its upper stage. The much larger *Ares V* had its problems too, necessitating a wider range of development tasks to add to the burgeoning cost.

One major competitor to the NASA government plan for a heavy-lift launch vehicle was the *DIRECT* concept put together by a number of NASA engineers and scientists as an alternative to *Constellation*. They wanted to offer a low-cost solution to the Moon programme and put together a valid proposal maximising the use of Shuttle elements in a different way and at less cost. The name came from the 'direct' transition from Shuttle to a new and diverse range of hardware including use of the Shuttle configuration for lifting large payloads and heavy cargo in a pod replacing the winged Orbiter.

The rocket launch vehicles embraced by *DIRECT* were diverse and broadly represented the best that could be harvested from existing technology. Frustration among some NASA engineers had already fuelled a trickle of migration across to the commercial sector, accompanied by a larger group from several major rocket builders serving contracts to NASA and the Department of Defense.

Right: As head of manned space flight from 2005 to 2019, William Gerstenmaier guided NASA's deep-space plans through their most difficult years. He is now VP of Build and Fly Reliability at SpaceX. (Arstechnica)

Below: As NASA looked to cheaper objectives for manned space flight, concerns over asteroid impacts (bolides) led to studies of ways to deflect them. Shown here are impact events recorded between 1994 and 2013 for objects 120m (360ft) in size. (NASA-JPL)

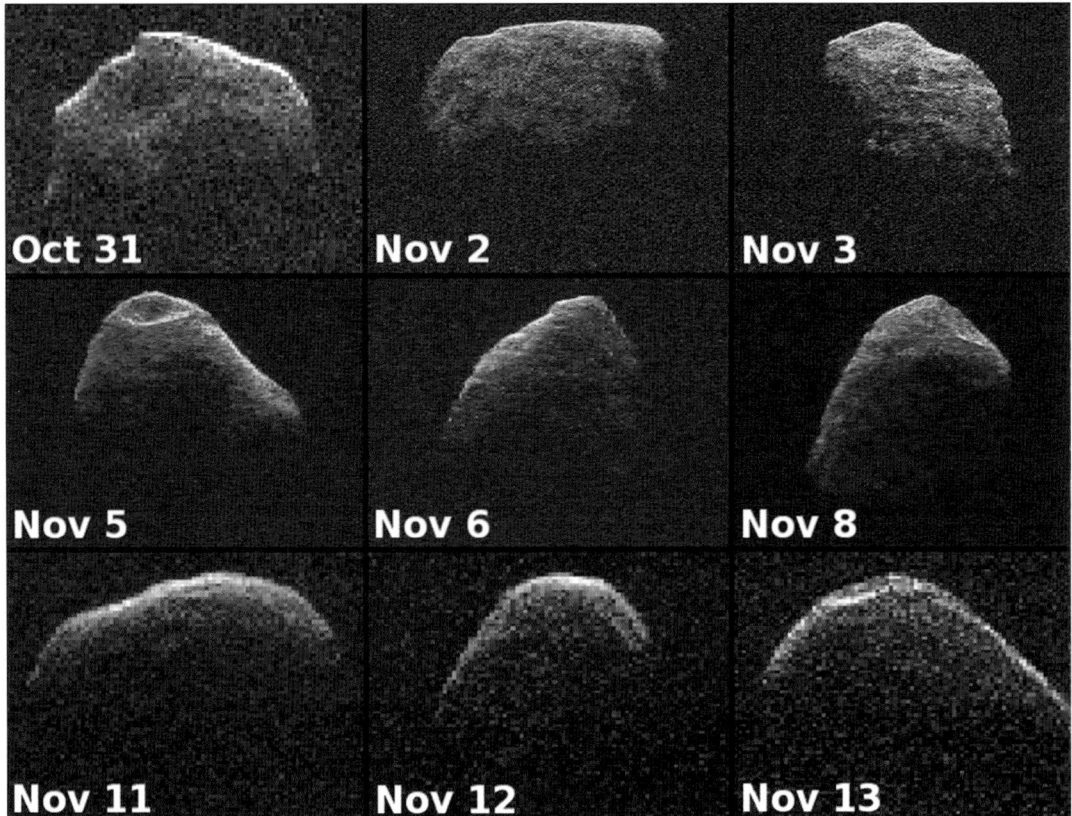

An asteroid 1.4km (0.87 miles) in diameter, observed by radar at a distance of 6.5 million km (4 million miles) during a close approach to Earth in 2012. (NASA)

The competitive edge in the commercial world of 'New Space' rocket builders began to grow and show greater potential than the less flexible 'Old Space' work of the government and established industry.

DIRECT was a highly credible approach with the potential to provide a broad and expanding set of hardware for a wide range of mission options. It was, however, outside the establishment and adrift from the more focused work being conducted by the entrepreneurs such as Elon Musk, who as CEO of SpaceX was fast becoming the poster boy of New Space and all that it promised. The NASA establishment reacted negatively to *DIRECT* and it never got the airing it deserved.

As the Augustine Commission pursued its detailed analysis of the *Constellation* programme, it held hearings at various NASA facilities and interviewed a wide range of industrial partners. A grave sense of foreboding descended upon these groups, which debated the consequences of an outright cancellation, already being considered a probability. In the five years since President Bush had announced his new Vision for Space Exploration, several astronauts left the Shuttle programme to accept senior management positions with industrial contractors serving the space industry.

Among those was former astronaut Dick Covey, who had left NASA in 1994 to join United Space Alliance (USA) as CEO. He spoke out volubly in defence of a robust national space programme defined by work already underway on *Constellation*. USA had the government contract to manage and coordinate the multitude of Shuttle contracts and integrate a smooth support infrastructure for that programme. In 2008, it had received a similar contract for work supporting *Constellation*.

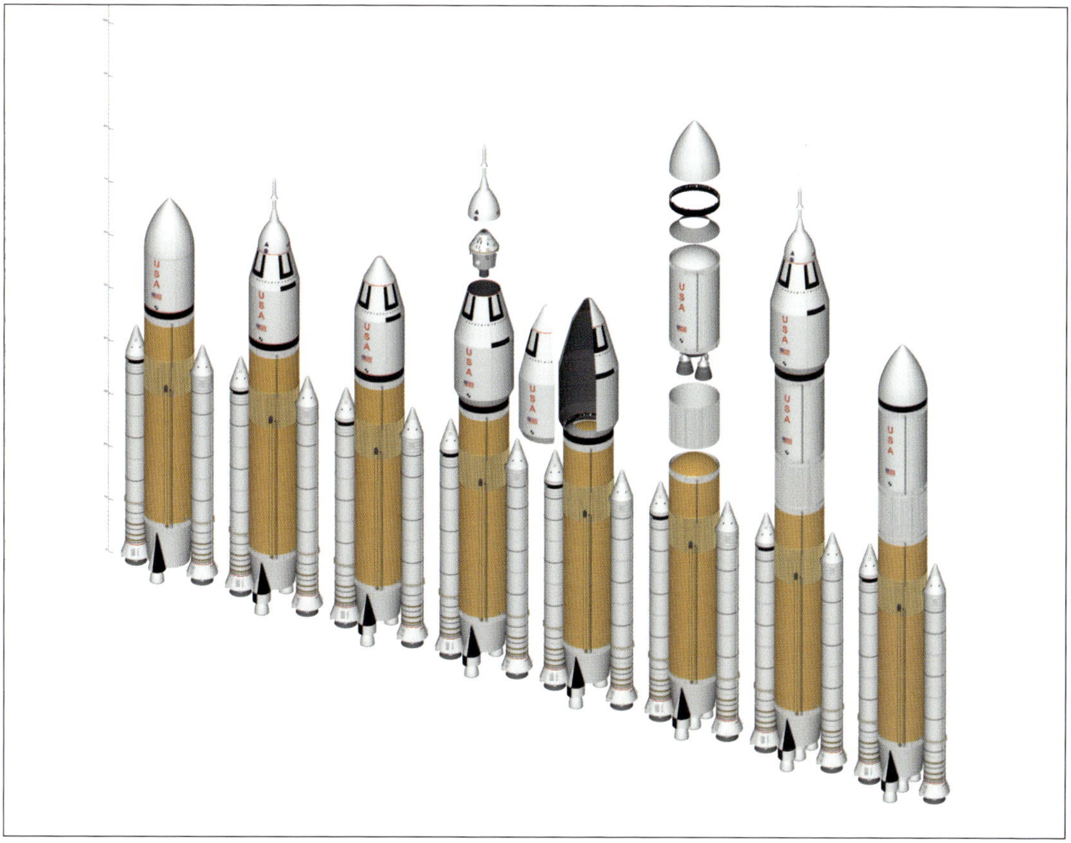

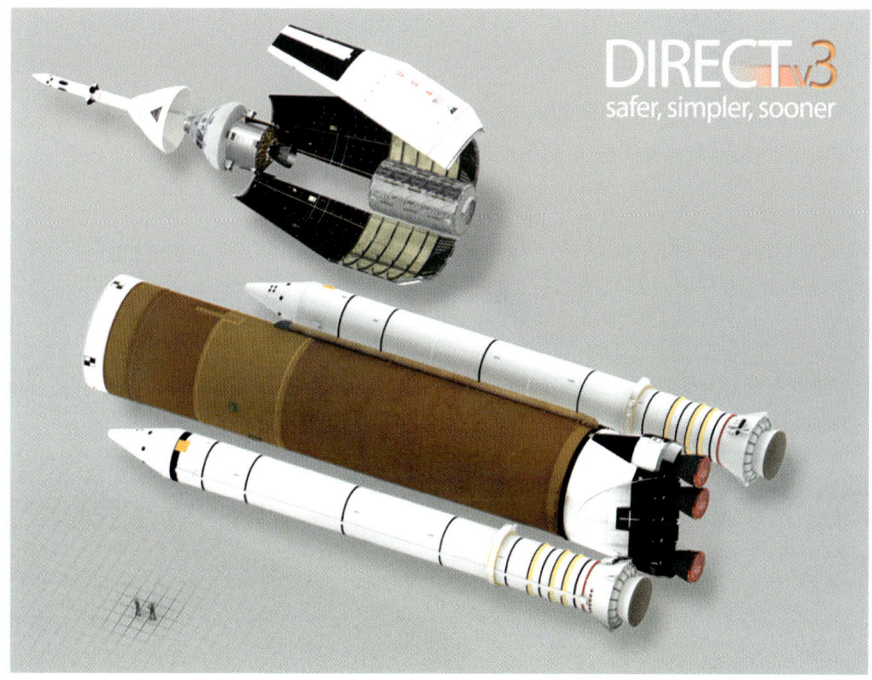

Above: Advocates of the DIRECT approach set out a range of *Jupiter* launch concepts, which they asserted were simpler and less costly than the *Constellation* vehicles. (Kraisee)

Right: The *Jupiter 130* configuration proposed as an alternative to the *Ares* vehicles baselined by NASA. (Kraisee)

Close to the time of completion of the report from the Augustine Commission, in August 2009, the Government Accountability Office (GAO) released a grim oversight of NASA's *Constellation* programme. It cited 464 high-risk areas of the *Orion* and *Ares I* projects, also identifying several key technologies that were either way behind schedule or displaying considerable challenges with few apparent solutions. It also highlighted an increase in the value of *Constellation* development contracts from $7.2 billion to $10.2 billion between 2007 and 2009, due to finding solutions for design and technology challenges.

The intended date of the first crewed flight for *Orion* had shifted to March 2015. Evidence brought before the GAO also revealed that NASA had moved to a two-block approach. An initial Block A version would conduct flights to and from the ISS, followed by flights with the more advanced

Left: *Jupiter 232* utilised the powerful RS-68 rocket motor in the core stage, two J-2 motors in the upper stage, and four-segment Solid Rocket Boosters. (Kraisee)

Below: All *Jupiter* configurations used hardware from the Shuttle programme and reconfigured it for deep-space missions using a ballistic capsule. (Kraisee)

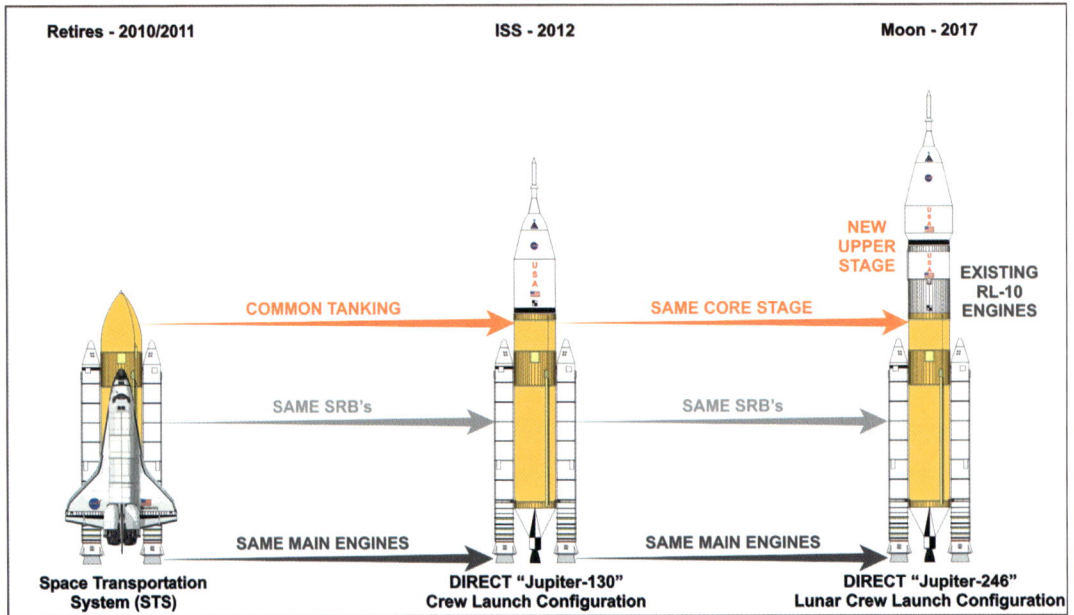

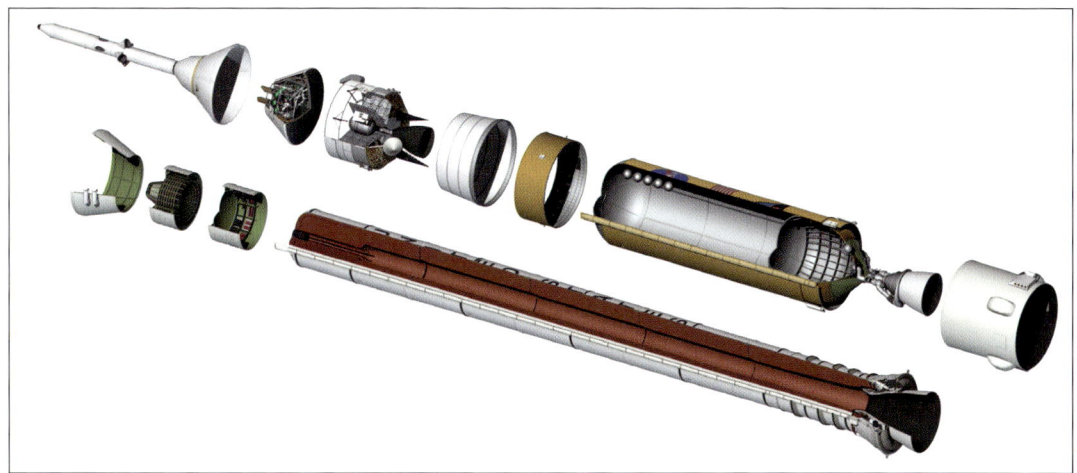

Above: The configuration of the *Ares I* launch vehicle, which had been planned as the rocket for placing payloads in Earth orbit. Although under threat, the *Ares I-X* test launch was allowed to go ahead. (NASA)

Right: Had *Ares I* been adopted for production it would have carried the CEV, later designated MPCV and then *Orion*, but it never flew with active upper elements. (NASA)

Block B for lunar missions. NASA claimed that the Earth-orbit flights were essential for proving the technologies essential for Moon missions. In reality, it could not afford to develop all the technologies simultaneously.

An uncertain direction

When the Augustine Commission reported its findings in October 2009, it reached several conclusions, foremost among was a judgement that the existing *Constellation* programme was unaffordable and too ambitious for the resources at hand. It endorsed ending Shuttle flights as planned in 2010 or 2011 after final assembly of the ISS, retaining operations with that facility until at least 2020 to reap benefits from the investment.

The Commission was ambiguous about how to maintain crew flights and cargo deliveries to the ISS after 2011 and acknowledged that there would be a gap of several years before the US could replace the Shuttle with a domestic system. It endorsed the opportunity to support development of a commercial crew vehicle for that task and this appeared to garner a lot of political opinion at the White House, but few in Congress were alert to the opportunities. The Commission was light on what to do next and, mindful of the decisions yet to be made by the Obama administration, it presented options rather than conclusive recommendations.

Meanwhile, as the Augustine team were taking the pulse of the nation through local town hall meetings, interviewing the great and the good in the aerospace fraternity, conducting a thorough

Above left: *Ares I-X* is rolled out to the pad on the Mobile Launch Platform (MLP). (NASA)

Above right: *Ares I-X* on the LC-39B pad at the Kennedy Space Center facility, which was still configured for Shuttle operations. (NASA)

survey of the state of the industry and opening Facebook pages for public viewpoints, NASA was testing *Ares I*. Under the programme plan for getting the Earth-orbit launcher tested, five flights would take place prior to the first manned *Orion* mission in 2015. *Ares I-X* would test the solid rocket booster stage, lifting an inert upper stage and a mock-up of an *Orion* capsule and its Launch Abort System. *Ares I-Y* would follow in late 2012 with a live upper stage, followed by three unmanned test launches of the fully configured *Ares I/Orion* in 2013 and 2014.

Ares I-X had been scheduled to fly on 27 October 2009, but weather and some minor technical issues postponed the attempt to the following morning. The configuration was with a four-segment booster; the fifth segment, which would be standard for an operational *Ares I*, was a mass simulator and not active on the I-X flight. Launch occurred at 11.30hrs EST on 28 October from Launch Complex 39B (LC-39B) at the Kennedy Space Center.

The stage fired for almost two minutes, propelling the rocket to a maximum altitude of about 45km (28 miles). Four seconds later, it separated from the upper elements, which tumbled away and were not recovered. The stage itself landed 240km (150 miles) out in the Atlantic Ocean. In further evidence of how far behind schedule the *Ares* programme was, *Constellation* programme manager Jeff Hanley requested cancellation of the *Ares I-Y* flight because the J-2 engine for the second stage would not be ready in time.

As noted, during the second half of 2009 there was an overriding fear that the Obama administration would cancel the entire programme and use the Augustine report to justify it. But, until ordered to the contrary, NASA kept going and began searching for possible paths it could follow should the Bush plan be axed. It found three potential fall-back options. One was the use of *Orion* for supporting the assembly of large observatory-class structures in near-Earth space and at more distant destinations. Another was to develop a very large Earth-observing telescope for extended scientific data-gathering of the planet and its atmosphere. A third possibility took into account the ongoing concerns about asteroids hitting Earth, by which *Orion* could form part of an expanded programme to support studies about Near-Earth Objects (NEOs), asteroid-size bodies which could threaten life on Earth. Robotic precursors of flights to these objects with the *Orion* spacecraft would, said NASA, make an effective use of the new vehicle.

A major revision could replace *Constellation* with a manned flight to visit Phobos, one of the moons of Mars. This could be achieved using a Crew Transfer Vehicle (CTV) to carry an expeditionary crew from Earth to an orbit about Mars. There, it could rendezvous with the tiny moon while its crew conducted spacewalks to investigate the surface and collect samples. This would be a major application of a new deep-space objective without requiring surface landers such as *Altair*. It would minimise the amount of new hardware and cost considerably less than the original plan for bases on the Moon and Mars surface visits. But why Phobos? It was a possibility revealed through a search for the most that could be achieved with the least hardware at the lowest cost.

An irregularly shaped body, Phobos is small, measuring 22 x 27 x 18km (13.6 x 16.7 x 11 miles) and orbiting Mars 6,000km (3,700 miles) above its surface. Its most prominent feature is the crater 'Stickney'. With a diameter of 9km (5.6 miles), Stickney was probably formed by an impact so great that it must have come close to shattering the moon itself. Creating a blanket of ejected material that rained across the surface, giant boulders scarred the surface and left prominent grooves running across the tiny moon.

Phobos is one of two moons of Mars, the other, Deimos, being considerably smaller with a surface area only one-third that of its inner neighbour. Moreover, Deimos orbits the planet at a distance of 23,460km (14,580 miles). Both moons lie almost exactly in the equatorial plane of Mars itself. Formed perhaps from large accumulations of dust, particles and small rocks, Phobos is slowly falling towards Mars and, in less than 50 million years, will impact the planet, or break up due to gravitational tidal forces. Until then, it provides a useful vantage point from where to conduct highly detailed surveys of Mars.

Ares I-X lifts off for a ballistic flight at 10:30hrs EST on 28 October 2009 with only four of the five-segment SRBs live, the remaining fifth segment and upper elements being dummies or mock-ups. (NASA)

Above: As *Ares I-X* roars skyward, Shuttle Orbiter *Atlantis* reposes on LC-39A, ready for the STS129 mission, which will begin on 16 November 2009. Only six more Shuttle missions remain. (NASA)

Left: *Ares I-X* passes through the region of maximum aerodynamic pressure, squeezing moisture from the atmosphere. (NASA)

While criticism of the Constellation Moon programme increased, NASA again began to look at using deep-space exploration to sample an asteroid, to better understand potential threats from Near-Earth Objects (NEOs). (NASA)

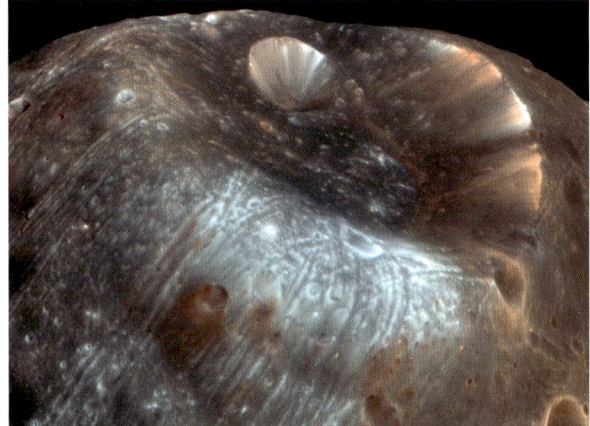

Above left: Some scientists wanted to engage with deep-space manned flights in the exploration of Mars's moon Phobos, seen here with extensive scarring and a deep crater called 'Stickney' at bottom right. (NASA-JPL)

Above right: With a diameter of 9km (5.6 miles), Stickney is evidence of a large impact on Phobos, a place geologists seek to explore. (NASA-JPL)

Right: Another view of Stickney on Phobos, a crater that dates close to the origin of the moon itself and holds secrets to the origin of both Phobos and Deimos. (NASA)

While there were many advantages to conducting a mission to Phobos – even Deimos, although with strikingly few stand-out benefits – there were significant challenges, one of which was the radiation factor. While a reconnaissance of Mars and a site survey of Phobos would be exciting, and carry much public attraction as it put people down on the surface of a second extra-terrestrial body, a surface landing could rest the crew for 18 months before they embarked on the return journey. A mission to Phobos would last at least 18 months in two-way journey time alone, exposing the crew to radiation over an extended and unbroken period. Although this was surmountable, as noted earlier, NASA had aired its concerns when it wrote DRM-6 for a manned Mars mission.

Like a phoenix

For a year after assuming office, the Obama administration remained silent about its intentions regarding NASA, future space policy and, in particular, the *Constellation* programme. It declared its hand on Monday 1 February 2010, seven years to the day after the loss of *Columbia*, an event that had triggered a deep look at options for a post-Shuttle future and a train of events that led to the *Constellation* programme embracing *Orion*, *Ares I*, *Ares V* and *Altair*.

Without announcing it formally to NASA personnel, Obama issued the formal budget proposal for Fiscal Year 2011, a 12-month period beginning on 1 October 2010, in which there was no money at

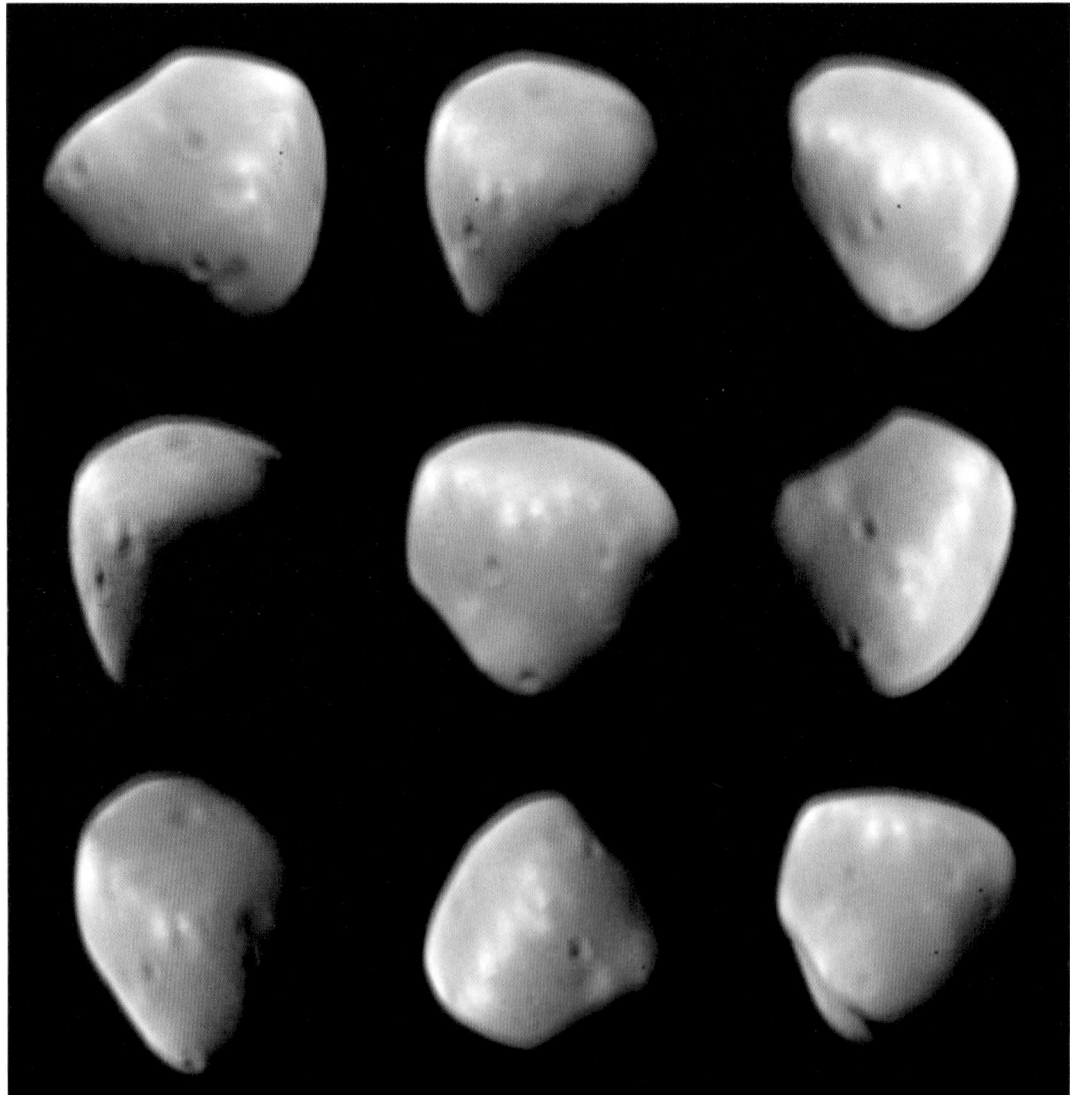

Orbiting farther out than Phobos, as seen in this range of views from different angles, Deimos is a small potato-shaped moon with a mean radius of 6.2km (3.8 miles), and unlikely to receive a visit from astronauts. (NASA)

all for *Constellation*. Employees and the multitude of workers on the *Constellation* programme heard it first in the formal budget announcement. There have been many fiery moments in Congress over budget issues, but the furore that broke out when senators and representatives heard the news was seminal. Without any real consultation, the government had torn up plans matured over many years and unilaterally rewritten the agenda for America's future in space.

Unlike constitutional democracies such as that in the United Kingdom, where the government has a majority in the legislature over all other parties combined and can unilaterally decide policy and pass it into law, decisions over government funding made in a representative democracy such as that of the United States require a broader debate and a greater legislative scrutiny. To quote the adage: the President proposes but Congress disposes!

Accordingly, having worked with largely bipartisan support to review, approve and fund *Constellation*, the manner in which it was proposed to end the programme left some bitterness. It also reflected badly on an incoming NASA Administrator who, without experience in the political machinations of Washington, DC, was left to explain the government's position to an angry legislature. But it was still up to Congress to handle the dramatic shift and resurrect a human space flight plan for the United States.

Politicians were aware of the national outcry at this decision which, according to polls, the majority of the general public saw as a flawed judgement. The Augustine team had conducted a national survey in which two-thirds of Americans believed human space flight benefited the country, was good for the economy and served as a stimulus for STEM subjects and higher learning for the next generation of students.

Right: Robotic cargo-carriers could have been utilised for exploring Phobos as well as the surface of Mars. (NASA)

Below: A further development of the commercial cargo programme for serving the International Space Station was commercial crew contracts for carrying astronauts, as with the SpaceX Crew Dragon. (G. De Chiara)

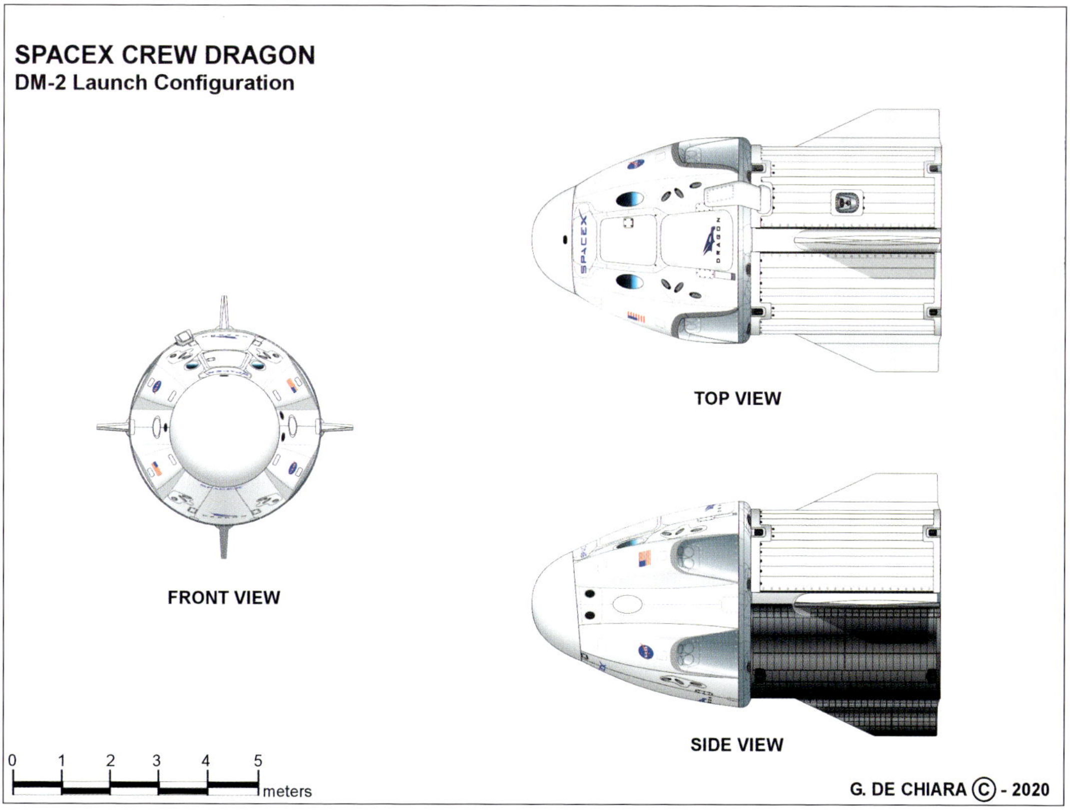

SPACEX CREW DRAGON
DM-2 Launch Configuration

TOP VIEW

FRONT VIEW

SIDE VIEW

G. DE CHIARA © - 2020

Crew Dragon was launched on top of a *Falcon 9* rocket, seen here in the assembly facility at the Kennedy Space Center. (SpaceX)

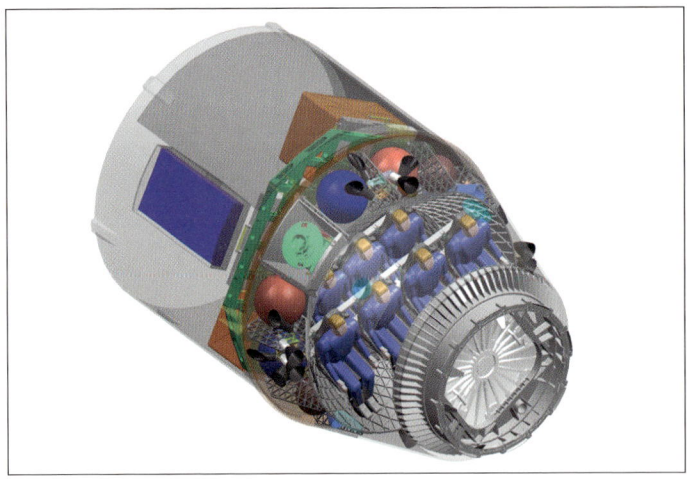

Designed for carrying up to seven crew members for flights to the International Space Station, Crew Dragon accommodates four astronauts. (SpaceX)

However, as reviewed earlier, there had been many concerns about the delays and cost overruns that plagued *Constellation*, but there was nothing to immediately take its place. Even the use of *Orion* to service the ISS with crew and cargo was in question, since there was now no launch vehicle to send it into space. But that would change, and there were aspects to Obama's new budget that would give direction to a future more certain and assured than could have been effected through *Constellation*. Not least was the money requested for supporting new technology initiatives through commercial cargo and crew programmes.

On the day the world heard that *Constellation* had been cancelled outright, NASA issued a tiny amount of seed money to Sierra Nevada, Boeing, United Launch Alliance, Blue Origin and Paragon Space Development Corporation for technology development toward a commercial crew vehicle to carry astronauts back and forth to the ISS. These were low-level contracts worth a total of $50 million. The real start-up fund was announced on 18 April 2011, when Blue Origin, Sierra Nevada, SpaceX and Boeing received a total of $270 million to develop vehicles to carry crew for the station.

The Shuttle would still be retired in 2011, but the possibility of employing commercial carriers from 2015 began to look increasingly at risk when Congress chafed at providing those companies with money to develop systems in competition with government programmes already spending taxpayer

Access to the Crew Dragon is via a bespoke walkway at the launch complex. (SpaceX)

The Crew Dragon capsule in preparation for the second demonstration flight. (SpaceX)

funds. And that was not all that Congress was concerned about. By cancelling *Constellation* and hoping for the eventual availability of commercial carriers, the US would rely on Russian *Soyuz* spacecraft to carry its astronauts to the station. Over time, that would cost the US taxpayer more than $4 billion, as the Russians increased seat prices from an initial $21.3 million to more than $90 million by the time the last seat slots were purchased in 2021. Added to this was approximately $8.6 billion paid to the US commercial contenders in development money and fees for launch services awarded by 2022. It had taken almost a decade for SpaceX to get its *Crew Dragon* developed, test-flown, certified and accepted by NASA for contracted flights carrying its astronauts back and forth instead of relying on Russian spacecraft and their rockets.

The need to buy Russian seats on *Soyuz* spacecraft lasted far longer than anticipated because Congress was reluctant to award the commercial companies the money they needed to fast-track the new spacecraft. The companies themselves had little inclination to do more than top-up government subsidies, and while Congress consistently awarded less money for that than the government requested each year, it took a lot longer to transition to US-owned and operated spacecraft.

Above: The Crew Dragon has a spacious interior with foldaway seats and a fold-out instrument panel, a far cry from cramped confines of early manned space vehicles. (SpaceX)

Left: Crew Dragon comprises a re-entry vehicle for the crew and a service module providing power, propulsion, environmental control and communications. (NASA)

In all, to date, almost $13 billion of US taxpayers' money has been spent to keep first Russia and now commercial contractors flying US astronauts to the space station. In addition, NASA has spent about $12 billion paying SpaceX and Orbital Sciences to fly cargo to and from the ISS. That money has gone into helping develop the rockets and the spacecraft now used to sell services to NASA. The transfer of government-owned services to subsidise the development of a new commercial space industry had brought both advocates and critics.

It was within this emerging but seismic shift that Congress began to examine the future for US space policy after the cancellation of *Constellation*. The basic policy plan presented by the President extended the life of the ISS to 2020, with crew and logistical services contracted out to commercial companies. In return, Congress agreed to cancel *Constellation*, but in an authorisation act dated 11 October 2010, it redirected NASA to other deep-space objectives more in line with budgetary expectations and new capabilities.

To do that, it ordered NASA to develop a launch vehicle capable of placing 70 metric tonnes (154,350lb) in low Earth orbit but with expansion to 130 tonnes (286,650lb) over time. This was less than *Ares V* but within reach of existing technology, removing new and advanced technologies that had challenged the *Constellation* programme. Congress directed NASA to develop a Multi-Purpose Crew Vehicle (MPCV) with an initial flight test in 2017. Clearly, the existing *Orion* vehicle, on which so much time and money had been spent, was the obvious choice.

The analysis and planning for a new heavy-lift vehicle took place in late 2010 and the concept review for what was now called the Space Launch System (SLS) ran through March 2011. By November the systems requirements review for the SLS had been completed, with a very wide range of Shuttle-derived systems forming a filter for optimised configurations. The SLS was being designed without a specific mission requirement in mind, but it was based around the *Orion* spacecraft and deep-space destinations yet to be determined.

Uppermost in potential missions was the NEO flight to visit an asteroid. For that, the SLS would be ideal. As determined, the giant rocket would have a total height of 96.92m (318ft) and comprise a

Right: Crew Dragon approaches the International Space Station for its first demonstration mission in March 2019. (NASA)

Below: Crew Dragon docked with the International Space Station on the second demonstration mission in May 2020, marking the first manned flight of a commercial spacecraft. (SpaceX)

Crew Dragon at splashdown in August 2020 as it returns from the first manned commercial flight with two astronauts. (NASA)

Boeing is developing its CST-100 *Starliner*, which will join Crew Dragon in sharing rides to the International Space Station. (Boeing)

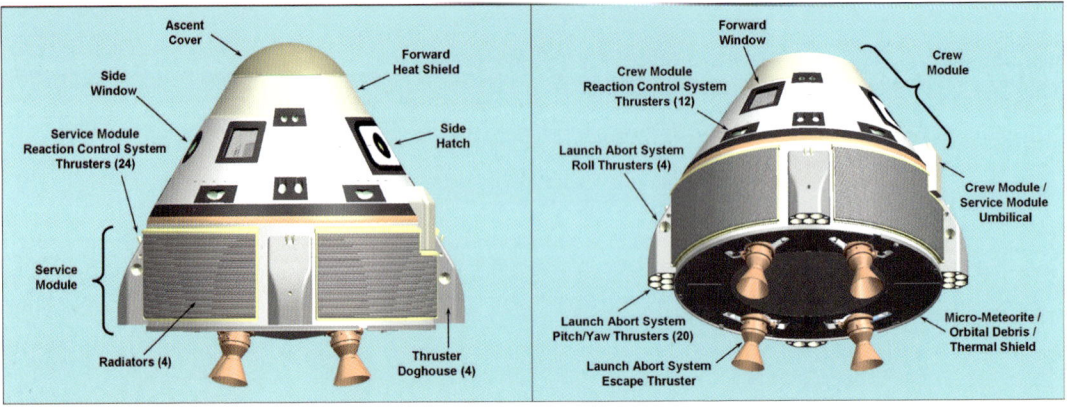

central cryogenic core stage, powered by four RS-25 Shuttle main engines at the base and flanked by two five-segment Shuttle SRBs, similar to the *Ares V* rocket. Initial flights were planned for 2017–21, with an upper stage powered by a J-2X rocket motor providing the 130-tonne orbit capability for the advanced version available from 2021.

A vision for the future

With approval for a new heavy-lift launcher and general acceptance of the *Orion* spacecraft as the MPCV, new opportunities began to emerge from the public and the private sectors. One of these lobbied for development of a permanent cislunar infrastructure, which meant a set of new technologies that could provide a permanent highway between the Earth and Moon. This was an idea first mooted in 1969, when

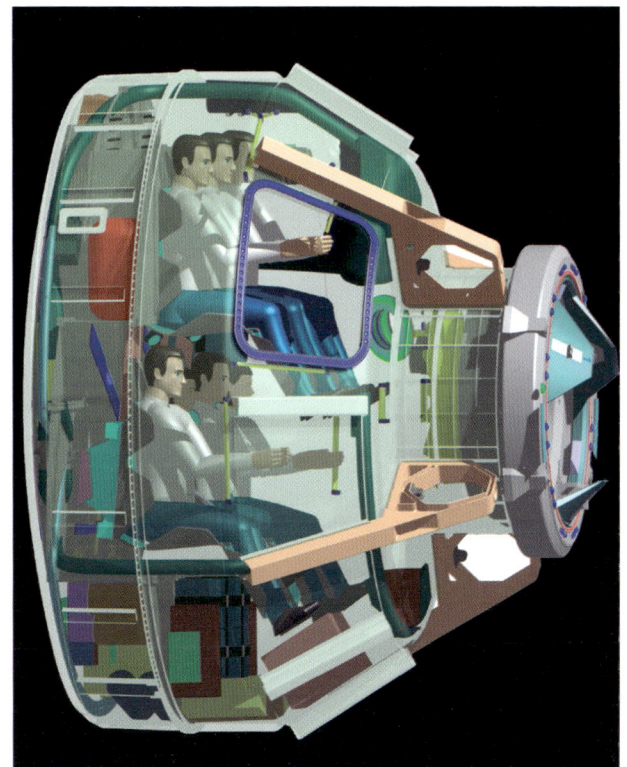

Right: *Starliner* will carry up to seven people and is designed to be recovered on land rather than water, the preferred mode for the SpaceX Dragon. (Boeing)

Below: Boeing's *Starliner* is carried into space on an *Atlas* launch vehicle, but will be compatible with other launchers as they become available. (Boeing)

Operations to the International Space Station could see the Boeing *Starliner* delivering astronauts to the facility from later in 2023. (Boeing)

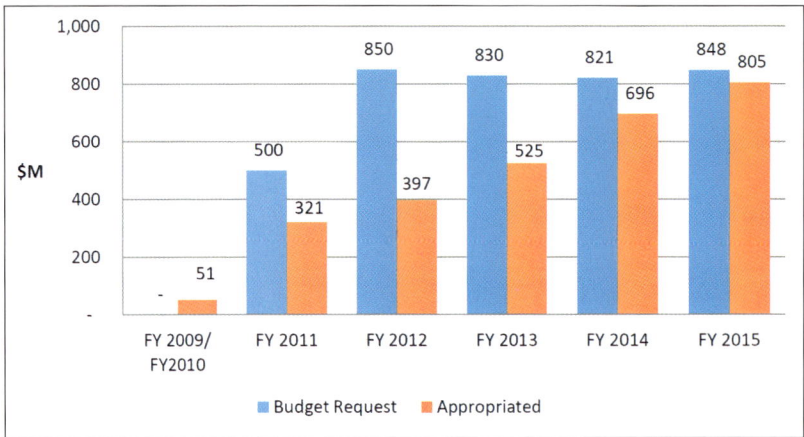

Delays to the commercial cargo programme were largely due to underfunding by Congress, which was reluctant to back private companies. (Author's collection)

NASA proposed a post-*Apollo* programme embracing a Shuttle, a permanently manned space station, a nuclear upper stage for moving cargo between Earth and the Moon and between the Earth-Moon system and Mars, and a Space Tug for moving satellites and spacecraft around between different orbits.

Forty years on, it was an idea resurrected from the ashes of the *Constellation* programme, which critics said was merely *Apollo* on steroids and did little to seriously revise the agenda towards a more permanent presence in deep space. The new call for a cislunar capability embraced the Earth and the Moon as a bi-planetary system from which could flow a wide range of capabilities. Unmanned robots had already informed scientists about resources on the Moon that could be exploited for use on Earth.

To service such a capacity for a new, space-based industrial revolution, advocates proposed a Gateway, a small space station in orbit around the Moon that would operate as a staging post for future activity. No longer thinking of the Moon as a place to get to from Earth, visionaries and futurists now regarded the Earth-Moon system as a single entity from which to depart for destinations across the solar system. A cislunar transportation system with refuelling depots for rockets, staging posts for launching toward Mars and for exploring and mining the Moon was within sight.

For any of this to happen, a revised national endeavour was essential and the availability of affordable hardware was an essential prerequisite. The single element of the original *Constellation* programme still in

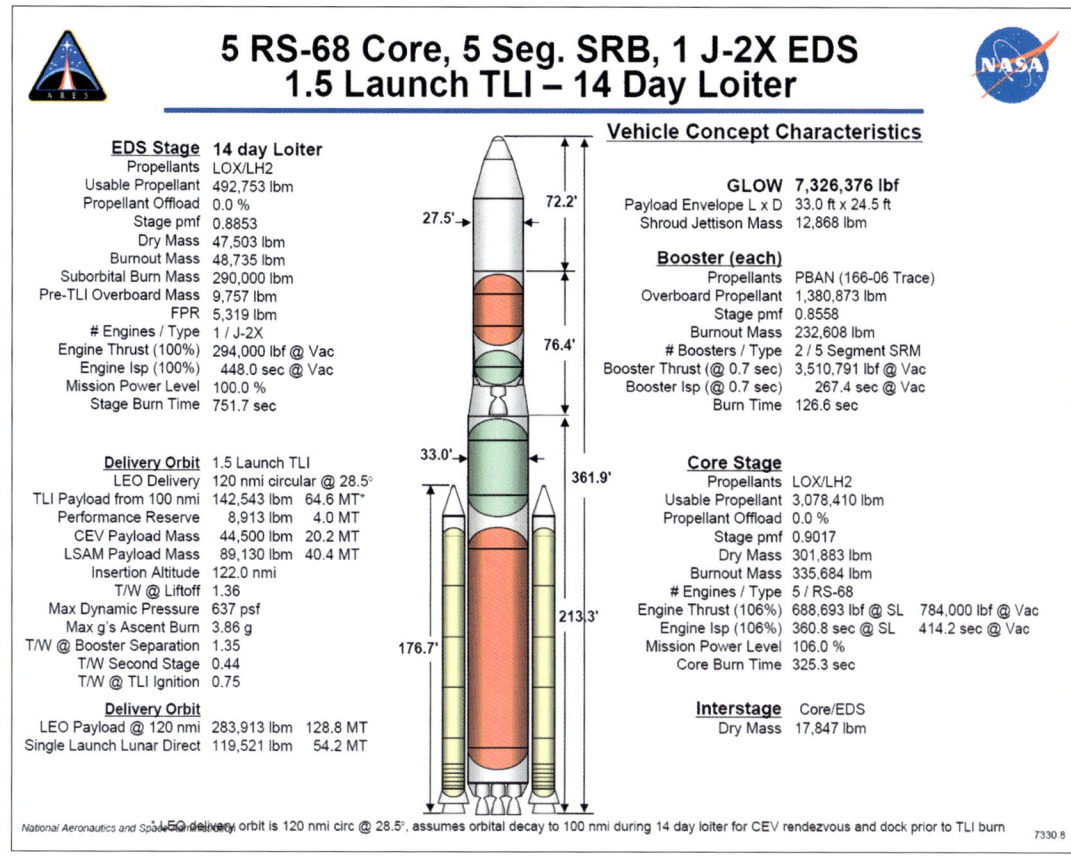

As displayed here, the fully evolved *Ares V* was retired as too expensive, with demanding technical challenges calling for big development budgets. It was cancelled along with the *Constellation* programme. (NASA)

play was the *Orion*, rebadged from CEV to MPCV. The crew module element was in better financial and technical shape than the service module, and the possibility of sharing that element with the Europeans began to appear as the only solution. Following informal discussions with the European Union that began in 2008, in May 2011 ESA formally declared that it was officially working to adapt the Automated Transfer Vehicle (ATV) into what would eventually be branded as the European Service Module (ESM).

In June 2012, Astrium, a European subsidiary of the EADS organisation, issued a couple of contracts for studies into the possibilities opened by adapting the ATV module into a service module for *Orion*. In November 2012, ESA member countries approved work on the ESM and production of a single flight module for the first *Orion* flight on an SLS began under an endorsement from NASA the following month. That work paid for Europe's access to the ISS between 2017 and 2020. On 17 November 2014, ESA awarded Airbus Defence and Space (ADS) a €390 million contract to develop and build the ESM in Bremen, Germany.

This was a seminal moment for Europe's aspirations to get aboard the post-Shuttle, NASA-led effort to explore deep-space destinations. ESA had a long and proud involvement with the US space programme, having built the pressurised *Spacelab* module carried inside the Shuttle payload bay for scientific research by international teams of astronauts. It had built the *Columbus* laboratory module for the International Space Station as a permanent part of that research facility and that was now integrated with *Orion*. It had also achieved an historic 'first', converting America's next manned spacecraft into a US-European vehicle.

Initially known as the National Launch System, the Space Launch System (SLS) concept was a close copy of the *Ares V* but with a less demanding performance requirement. (NASA)

With the SLS, NASA wanted to follow the concept of *Ares V* and envisaged it as following the same assembly procedure in the Vehicle Assembly Building (VAB) at the Kennedy Space Center. (NASA)

With the cancellation of the *Constellation* programme, NASA examined the possibility of a Lunar Gateway, a staging post that could use existing technology without breaking the budget. (NASA)

Partnerships on the International Space Station could be transferred to a Lunar Gateway where, as seen here, a Japanese module arrives to dock with the facility. (NASA)

Based on its experience with the *Spacelab* Shuttle module, Europe provided *Columbus*, one of the major laboratories for the International Space Station. (ESA)

Development of mission tasks for the new combination of SLS/*Orion* evolved rapidly and with full Congressional support. The Obama administration pressed for a commercial involvement in support for the ISS, but had little enthusiasm for a government-led, deep-space commitment. Nevertheless, working on earlier studies and supported by Congress, two prime objectives emerged: a flight into lunar orbit to prove deep-space capabilities with the 70-tonne capacity (Block 1) SLS and *Orion*, and a follow-on mission to a Near-Earth Asteroid (NEA) using the 130-tonne capacity (Block 2) SLS.

By 2014, the programme was working along that dual track with two more hardware elements: a Space Exploration Vehicle (SEV) and a Deep Space Habitat (DSH), the latter not unlike the Gateway proposal. The SEV would provide pressurised habitation for attaching to another vehicle or to an asteroid, while the DSH would provide a habitation for crew and their supplies on long-distance transits between locations, much like the Gateway concept but for extended journeys.

For really long journeys through space, NASA sought to develop a Solar Electric Propulsion (SEP) element, which would use light from the Sun to produce electrical current for a low-energy impulse. Electric rocket motors offer much promise because, although they have exceptionally low thrust, they operate on a continuous basis and so produce enormous speeds built up over days and weeks of sustained operation. In this way they can progressively accelerate a spacecraft to far higher speeds than can be obtained from a high-thrust chemical rocket motor, which quickly exhausts its propellant.

Electric thrusters are used on satellites for maintaining orbital position and have been used frequently for small, deep-space probes.

The two lunar-orbit and NEA reference missions would prepare the way for lunar landings through a series of incremental steps. The Moon mission would involve a Block I SLS sending *Orion* to what engineers call a Distant Retrograde Orbit (DRO), a path in the opposite direction to that in which the Moon orbits the Earth. *Apollo* used a retrograde orbit but the 'distant' component comes from being on a path that extends far beyond the Moon and circles it every two weeks. Such an orbit is stabilised by the gravitational masses of the two bodies (Earth and Moon) and requires very little propellant to adjust. *Apollo* orbits were close to the Moon, circling it every two hours, which requires a lot of propellant to remain stable.

A further refinement, known as a Near-Rectilinear Halo Orbit (NRHO), is highly efficient, can last from one to two weeks and is further stabilised by a balance of gravitational forces between the Sun, the Earth and the Moon. The NRHO path is highly suited to lunar-orbiting space stations such as the proposed Gateway, because it can remain without disturbance for long periods and does not require any significant amount of propellant to remain in position. While Distant Retrograde Orbits are nearly circular and far from the surface, NRHO paths are elliptical and can be set up to pass close to a specific point on the surface. This means that any spacecraft attached to such an orbiting facility can drop down from that low point and land.

By 2014, the future for human space flight appeared back on track, albeit with an uncertain set of goals. The hardware, however, was back in development and provided options. Nevertheless, from certainty came confusion once again as the objectives and mission targets were juggled before settling back once more on the original goal for *Constellation* – the Moon.

Attached to the International Space Station in 2008, *Columbus* provides workspace for scientists at the orbiting facility. (ESA)

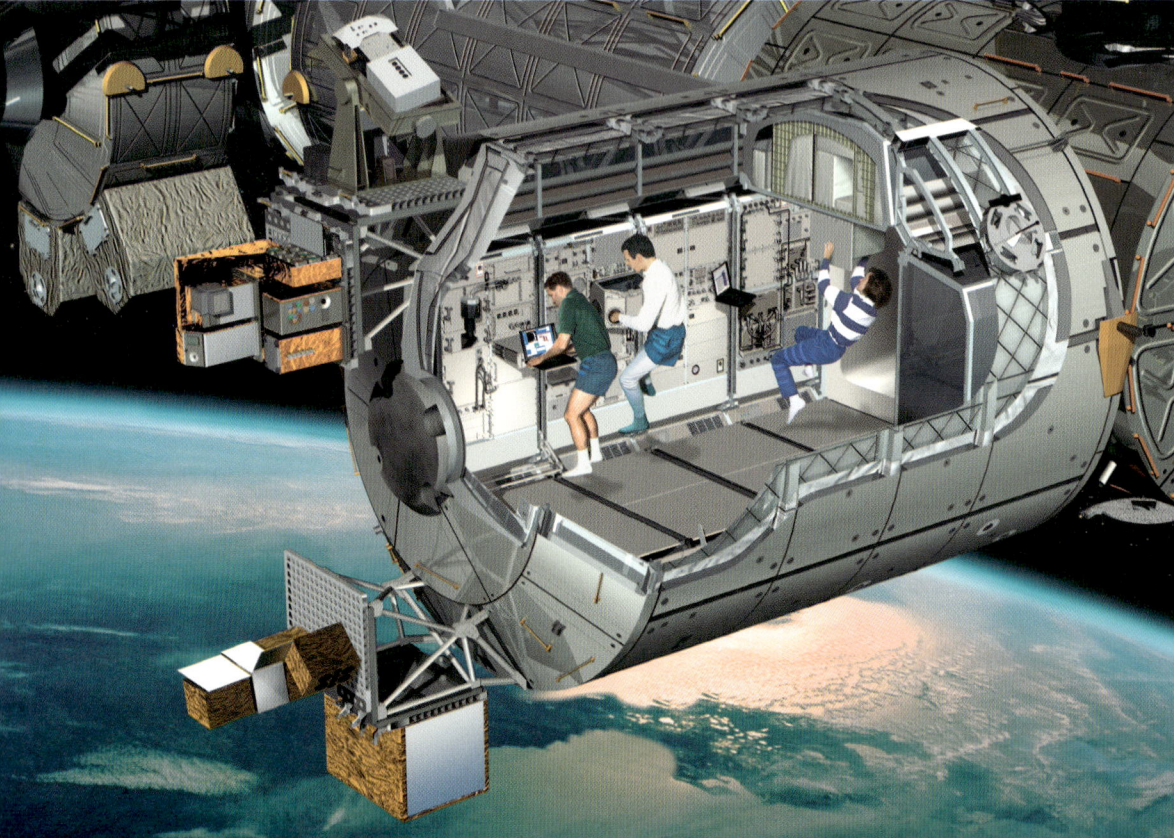

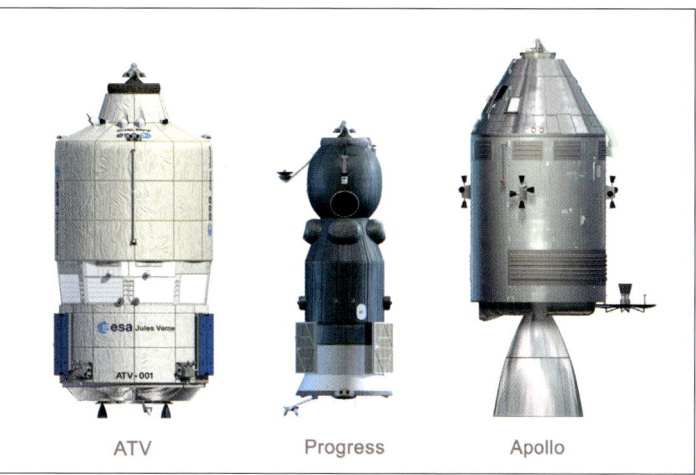

Right: Based on experience building *Spacelab* and *Columbus*, Europe produced the Automated Transfer Vehicle, seen here in scale with Russia's *Progress* cargo module and the Apollo spacecraft. (ESA)

Below: Escalation in costs for the MPCV (*Orion*) pushed NASA to cut a deal with ESA for producing the European Service Module (ESM) instead of *Orion*-builder Lockheed Martin. (ESA)

The core structural frame of the European Service Module. (ESA)

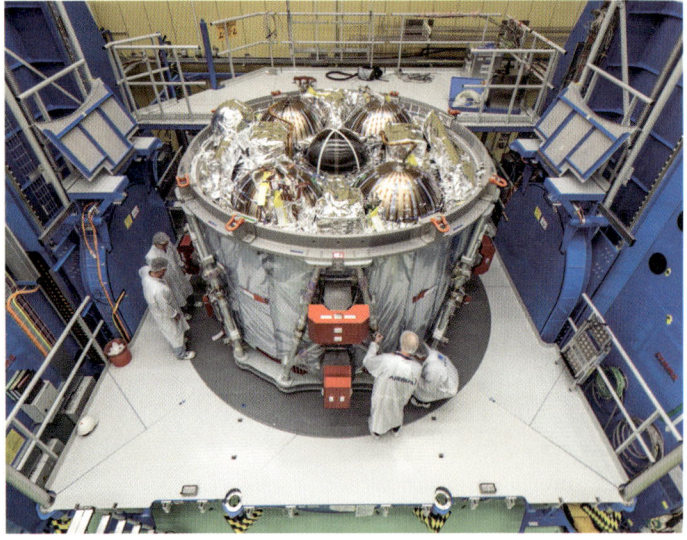

Four cylindrical propellant tanks with hemispherical end domes occupy much of the internal structure. (ESA)

NASA has been experimenting with Solar Electric Propulsion (SEP) for several decades, as a means of achieving access to distant targets in less time than with current rockets. (NASA)

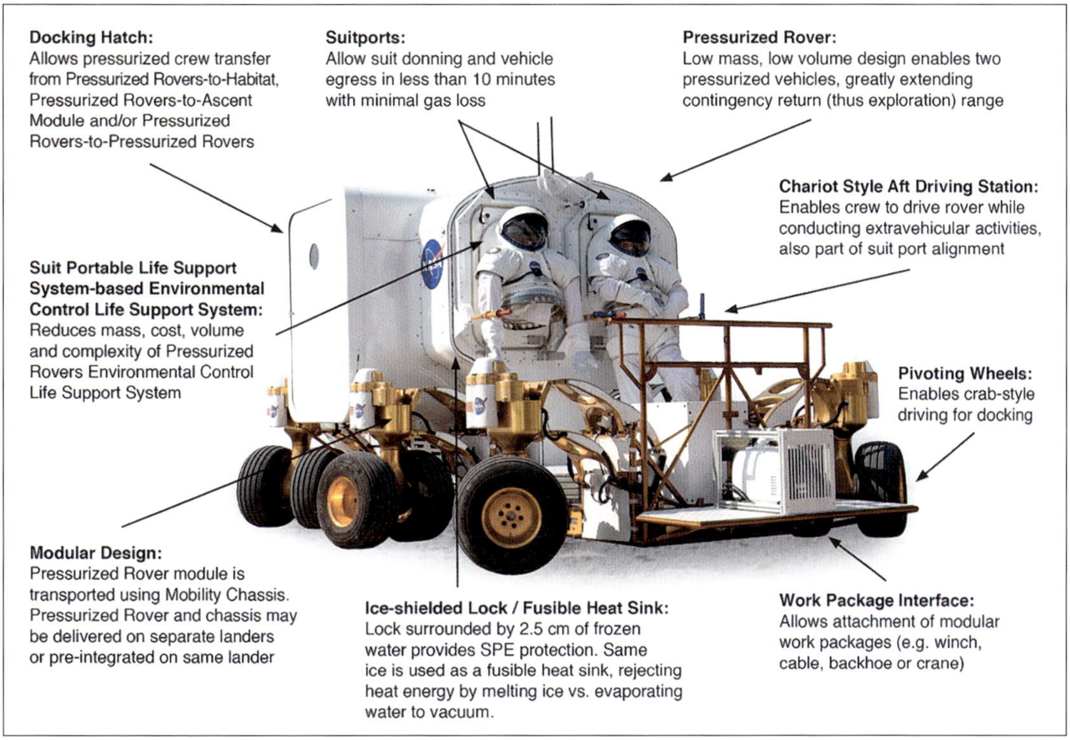

Docking Hatch:
Allows pressurized crew transfer from Pressurized Rovers-to-Habitat, Pressurized Rovers-to-Ascent Module and/or Pressurized Rovers-to-Pressurized Rovers
Suitports:
Allow suit donning and vehicle egress in less than 10 minutes with minimal gas loss
Pressurized Rover:
Low mass, low volume design enables two pressurized vehicles, greatly extending contingency return (thus exploration) range
Chariot Style Aft Driving Station:
Enables crew to drive rover while conducting extravehicular activities, also part of suit port alignment
Suit Portable Life Support System-based Environmental Control Life Support System:
Reduces mass, cost, volume and complexity of Pressurized Rovers Environmental Control Life Support System
Pivoting Wheels:
Enables crab-style driving for docking
Modular Design:
Pressurized Rover module is transported using Mobility Chassis. Pressurized Rover and chassis may be delivered on separate landers or pre-integrated on same lander
Ice-shielded Lock / Fusible Heat Sink:
Lock surrounded by 2.5 cm of frozen water provides SPE protection. Same ice is used as a fusible heat sink, rejecting heat energy by melting ice vs. evaporating water to vacuum.
Work Package Interface:
Allows attachment of modular work packages (e.g. winch, cable, backhoe or crane)

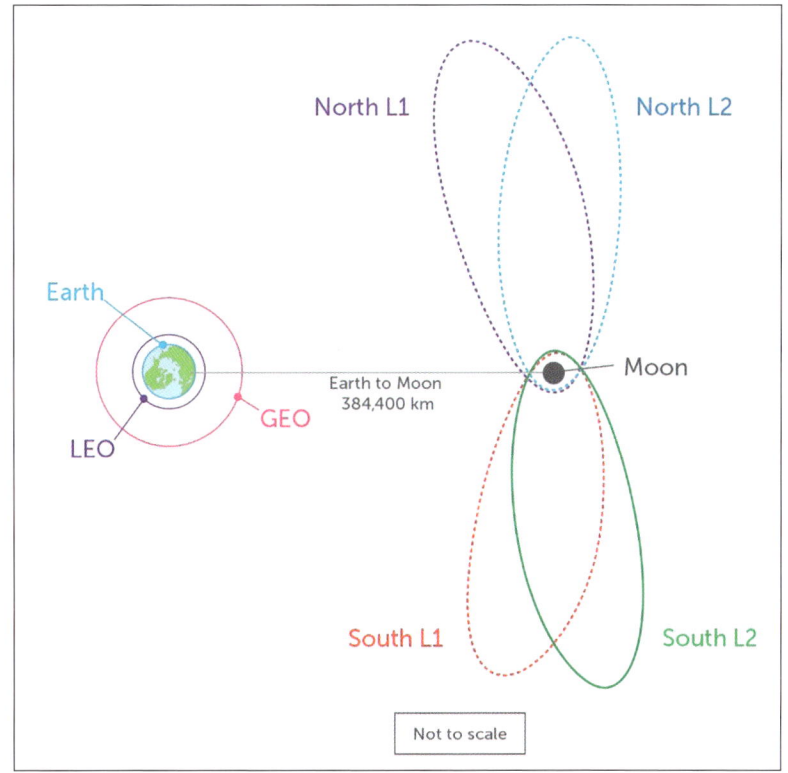

***Above*: Further planning for future Moon exploration included research on a Space Exploration Vehicle (SEV) capable of housing two astronauts for two weeks. (NASA)**

***Right*: The Near Rectilinear Halo Orbit (NRHO) is one in which the amount of energy to get into lunar orbit is very low, as is the periodic orbital adjustment to stay in place. (Author's collection)**

North L1
North L2
Earth
Moon
Earth to Moon
384,400 km
GEO
LEO
South L1
South L2
Not to scale

Chapter 3
Artemis Rising

In 2012, support began to gather for an ambitious expansion of the NEA concept to one in which a SEP-powered spacecraft would fly to an asteroid, attach a propulsion package and steer it into an orbit of the Moon from where astronauts could recover samples for return to Earth. This was named the Asteroid Retrieval Mission (ARM) and it gathered support and scepticism in equal measure. All along, NASA maintained that it would support the eventual goal of sending humans to Mars, although the critical path to achieve that by way of ARM was lost on many.

The plan was to start flight operations with Exploration Mission-1 (EM-1), which in 2012 was scheduled to see the first unmanned launch of the SLS on 17 December 2017. Prior to that, Exploration Flight Test-1 (EFT-1) was scheduled to fly the first *Orion* capsule for a test of its heat shield in a highly elliptical trajectory around Earth on a *Delta IV* Heavy in 2014. Following the flight of EM-1, a repeat (EM-2) scheduled no earlier than 2019 would hopefully qualify the SLS/*Orion* system for manned flight. But all that was for the future, and there would be significant changes to come.

Fabrication of the structural elements of the first *Orion* spacecraft (CM-001) for EFT-1 was completed at the Michoud Assembly Facility in June 2012. It was delivered to the Kennedy Space Center for final assembly in the Operations and Checkout (O&C) building. The *Delta IV* rocket was a powerful outgrowth of earlier launchers in this class, with a highly reliable success record. Sent aloft from Space Launch Complex 37B (SLC-37B) at Cape Canaveral, it would place *Orion* on a highly elliptical, two-orbit flight around the Earth on 5 December 2014.

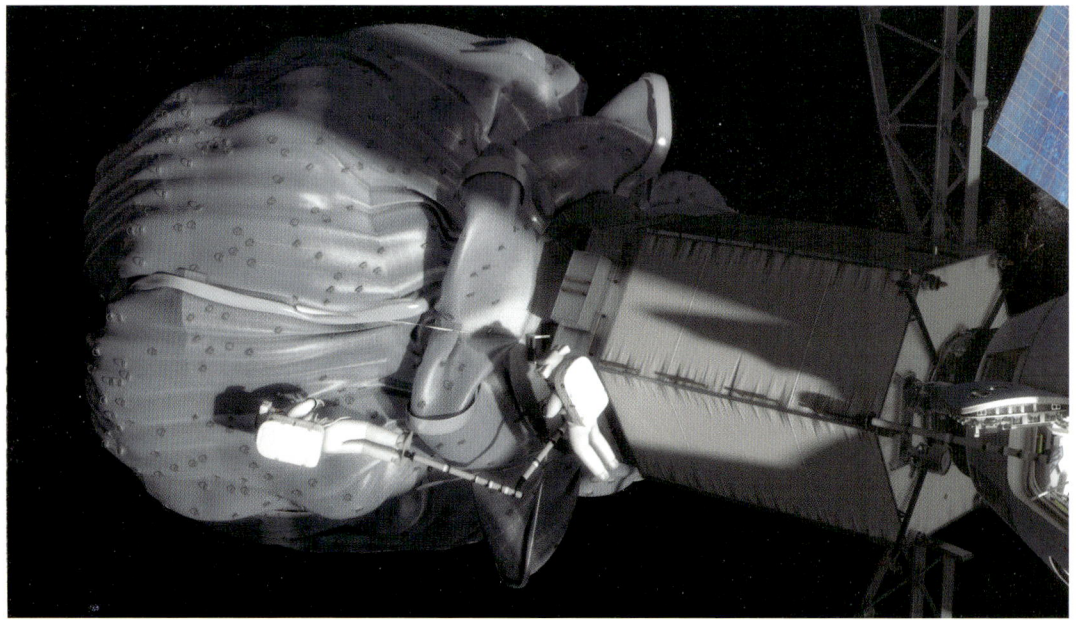

Searching for a meaningful mission, NASA turned to retrieving an asteroid by unmanned spacecraft for scavenging by astronauts flown on *Orion*. (NASA)

Above: Fabrication of CM-001 began in June 2012 at NASA's Michoud Assembly Facility. (NASA)

Right: The conical panels known as the backshell are fitted after most internal systems are installed in CM-001. (NASA)

At launch, the Heavy variant employed three massive liquid propellant stages aligned side-by-side for a total lift-off thrust of 9,474kN (2.13 million lb). The two flanking core stages acted as boosters and shut down at 3 minutes 56 seconds while the central stage throttled back, saving propellant and extending its burn time to 5 minutes 30 seconds. At that point, the core dropped away, leaving the remaining upper stage to ignite 16 seconds later. Powered by a single cryogenic RL-10 rocket motor developed in the late 1950s, the upper stage fired for 11 minutes 50 seconds, producing an orbit of 185 x 888km (115 x 552 miles).

The remainder of the stack drifted for 98 minutes. At the start of the second orbit, the J-2 ignited a second time, this time for 4 minutes 43 seconds, pushing the assembly to a peak altitude of 5,808km (3,609 miles), which it reached 3 hours 6 minutes after lift-off. Shortly thereafter, CM-001 separated from the now inert upper stage and the dummy service module, orientating itself for re-entry 4 hours 13 minutes 41 seconds after lift-off, where temperatures reached 2,200°C (4,000°F). Splashdown occurred at 4 hours 24 minutes 46 seconds at a gentle 27kph (17mph), some 965km (600 miles) off the coast of Baja California.

In most aspects the test flight was a success, but post-flight inspection revealed concerns about the heat shield, scaled up from that developed for *Apollo*. This prompted a complete redesign of the way the base shield was fabricated. Instead of a single fabricated structure, future heat shields would be made up from separate blocks built up in stages, allowing subtle changes to the way the shield reacts to its environment.

Unlike EFT-1, operational *Orion* flights would expose the shield to longer periods at lower temperatures in deep space and much higher temperatures on re-entry. These changes were added to existing plans for future heat shields to have a 3D woven quartz fabric in the six compression pads

The dummy structure replicating the service module for CM-001 on the EFT-1 mission is ready for the exterior closeout panels. (NASA)

The *Delta IV Heavy* is ready for launch at LC-37B, supporting the EFT-1 test flight for the first *Orion* Crew Vehicle. (NASA)

Delta IV Heavy's cryogenic upper stage and RL-10-B2 rocket motor as it would have appeared in orbit, still attached to the Orion spacecraft. (NASA)

Exploration Flight Test-1 launches at 07.05hrs EST on 5 December 2014, for a flight lasting 4 hours 24 minutes 46 seconds. (Hladuk)

that held the Crew Module to the European Service Module. In a further, albeit minor, shift, the entire backshell would be coated silver for thermal reflectivity, reducing internal temperatures.

By November 2015, the design, manufacture, assembly and test criteria for *Orion* were completed. The first piloted mission, Exploration Mission 2, was expected no later than April 2023. While technical milestones were completed successfully, over all the pace slowed and the cost grew. Still lacking a true commitment and a single, unifying goal, the objectives of the manned space programme lumbered along with only a vague idea as to its real value for cislunar transportation. All the while the SLS/*Orion* programme was languishing along in support of the Asteroid Redirect Mission.

In late 2016 the ARM programme was cancelled; the International Space Station had been extended to at least 2024 and the SLS/*Orion* role had been vaguely repurposed into support for an undefined 'cislunar capability'. By early 2017, a wave of change was blowing across Washington, DC, and the manned space programme was about to get turned on its head. Years of caution, delay, risk aversion and vacillation from an indifferent White House were about to be overthrown.

Trump opts for the Moon

Sworn in as US President on 20 January 2017, Donald Trump acted swiftly to restore a major, deep-space commitment to NASA's human space flight programme. On 11 December he issued Space Policy Directive No 1, directing NASA to 'enable human expansion across the solar system' and to return 'humans to the Moon for long-term exploration and utilization'.

To see the new initiative safely through Congress, Trump appointed Jim Bridenstine, a seasoned politico with long experience in the halls of government and the US Congress, head of NASA from 23 April 2018. A far-ranging mobilisation of NASA resources would push ahead with the development of commercial cargo and crew programmes, to send astronauts around the Moon on EM-2 in 2023, build a Lunar Gateway by 2025 and put US astronauts back on the Moon by 2028.

During 2019, the Trump administration re-badged the manned effort under the name *Artemis*, announcing on 26 March an acceleration that envisaged humans back on the Moon in 2024. Before that, EM-1 became *Artemis I* and remained the first unmanned test flight of the SLS and *Orion* in 2021.

Above: *Orion*'s Crew Module splashes down after a highly elliptical shakedown test, during which flaws were discovered in the heat shield. (NASA)

Left: The *Orion* spacecraft is recovered by the amphibious transport dock USS *Anchorage*. It is now on display at the Kennedy Space Center Visitor Complex. (US Navy)

With its redesigned heat shield, *Orion* would have no crew; it would conduct a four-week flight around the Moon to test out all systems and come home. EM-2, the first flight of *Orion* on an SLS with a four-person crew, would become *Artemis II*, now planned for 2022.

Under a decision to involve commercial companies in providing a Human Landing System (HLS), NASA-speak for Moon lander, the agency would rent the services of the contractor, which would own the hardware. This, said NASA, would conduct the first human flight to the Moon with *Artemis III* in 2024, four years earlier than the previous plan. In making the transitional shift, NASA was finally giving up hope of being able to develop a lander itself and opting to subsidise commercial providers

When Donald Trump became President in January 2017, he grasped the space programme as a means of 'making America great again' and would transform the manned programme. (NASA)

An experienced politico with a background on Capitol Hill, Jim Bridenstine became NASA Administrator in April 2018, championing a fast return to the Moon. (NASA)

building the spacecraft. In truth, major aerospace corporations had been building NASA's manned space vehicles all along, but the government owned those.

A second, third and fourth landing would fly over the next three successive years as *Artemis IV*, *V* and *VI*. A foundation habitat for sustaining a semi-permanent human presence on the lunar surface would follow on *Artemis VII* in 2028, with further flights to consolidate the base as *Artemis VIII* in 2029 and *Artemis IX* in 2030. That plan did not last long.

First, however, was the urgent need to select a contractor for the HLS. The bidding reference used an Advanced Exploration Lander (AEL) concept put together by the agency in 2018. That envisaged three stages weighing a total of 36,000–45,000kg (79,380–99,225lb), each launched separately and assembled at the Lunar Gateway. After launch from Earth, *Orion* would dock with the AEL, transfer crew to go down to the surface and ascend back at the end of the stay, with the bottom stage left on the Moon and the upper two stages reusable.

In December 2018, NASA declared its intent to invite bids, and on 14 February 2019, briefed industry on the requirements. The formal request went out in April and, by the 15 November deadline, five companies had responded, of which three were selected for further work and awarded separate contracts in April 2020. *Blue Origin* from Jeff Bezos and Dynetics were conventional lander

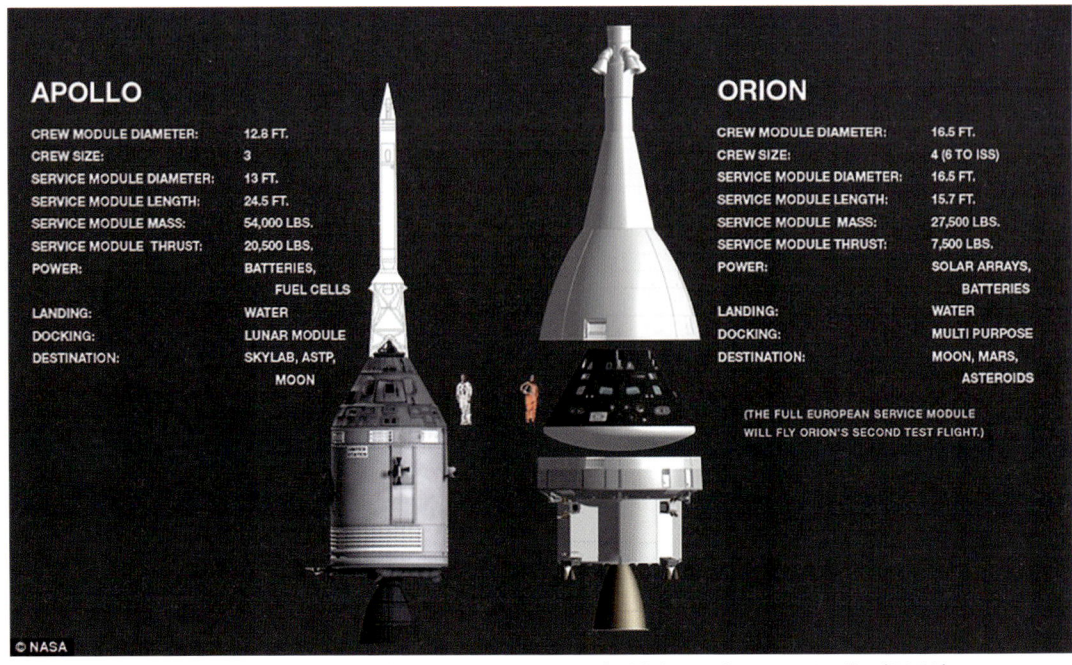

Above: With design details firming up, *Orion* was large compared with its predecessor Apollo. (NASA)

Below left: NASA put up its 2018 Advanced Exploration Lander as the template concept for commercial bidders on the Moon lander. (NASA)

Below right: The Moon lander imagined as it might look on the lunar surface, a conventional evolution of the *Apollo* Lunar Module. (NASA)

configurations but SpaceX had a novel proposal, adapting the upper stage of its massive *Starship* rocket, still under development, into a lunar lander.

Elon Musk had been working for several years on *Starship*, a two-stage, super-heavy launch vehicle with an upper element of the same name. It was this upper element that SpaceX offered as the basic design for HLS. The *Starship* upper stage is 50m (164ft) in height with a diameter of 9m (30ft), supporting three Raptor engines, each delivering a thrust of 2,300kN (520,000lb) from liquid oxygen and methane.

NASA judged the bids on three criteria: technical adequacy, management strength and price. It announced its decision on 16 April 2021, choosing SpaceX with the lowest bid at $2.94 billion versus $5.99 billion for Blue Origin and $9.08 billion for Dynetics. The two losers challenged the selection of SpaceX, which held up work on the SpaceX *Starship* programme until the end of the year, when a federal court upheld the decision. Responding to concerns from Congress, NASA ran a second set of contracts for an alternative concept, also requiring it to be a 'sustainable' landing system for flights beyond *Artemis III*. On 23 March 2022, NASA announced that it wanted the second-generation lander to be available by 2026, providing redundancy and competition. Moreover, it now required SpaceX to conduct a second unmanned test landing prior to signing off on the first manned *Artemis III* landing.

Each Moon mission will begin with the launch of a two-stage *Starship* into Earth orbit, where the upper element will remain as a propellant depot. Next, several *Tanker Starship* launches will rendezvous with the depot to fill it with propellant, which will be transferred to another *Starship* flight that brings up the HLS. Thus fuelled, the unmanned HLS will fire its rocket motors to go to the Moon and enter an elliptical NRHO path.

When all this has been completed, with the HLS checked out automatically, an SLS will launch *Orion* and a four-person crew from the Kennedy Space Center to rendezvous and dock with the HLS in lunar orbit. Two astronauts will transfer to the HLS, leaving two in *Orion*, and command it from the NRHO path to a closer orbit of the Moon. From there they will begin the descent to the surface. The crew will remain on the Moon for about a week and conduct several excursions outside the lander before returning to rendezvous and dock with *Orion* for return to Earth.

Rockets and spaceships

By the early 2020s NASA had upgraded the Block I SLS, now capable of sending a 95-metric tonne (209,000lb) payload to low-Earth orbit, upper stage included, raising to 27 tonnes (59,500lb) the

Blue Origin followed the NASA conceptualisation and proposed a close copy for its bid on the Human Landing System (HLS) contract. (Blue Origin)

A full-scale mock-up of the Blue Origin HLS proposal. (Blue Origin)

Left: Dynetics produced a mock-up of its proposed Moon lander and displayed a similarly conservative approach to that of Blue Origin. (Dynetics)

Below: The Dynetics HLS proposal could accommodate two astronauts for a week on the lunar surface and deploy solar panels for power. (Dynetics)

SpaceX offered a version of its *Starship* for the Moon lander bid and was successful in a contract awarded in April 2021. (SpaceX)

payload it could send to the Moon. But in addition to the Block 2 mooted earlier, the Block IB was envisaged with a low-Earth orbit payload of 105 tonnes (31,000lb), or 42 tonnes (92,500lb) to the Moon. To achieve this, the Interim Cryogenic Propulsion Stage (ICPS) of the Block I would be replaced with the more powerful Delta Cryogenic Second Stage (DCSS) from the *Delta IV* launch vehicle. When used on the SLS, it would be known as the Exploration Upper Stage (EUS).

Block I would fly the unmanned *Artemis I* mission to lunar orbit, the *Artemis II* manned flight around the Moon, and support the first landing on *Artemis III*. Block IB would be available for the following five missions. If developed, Block 2 would have new and more powerful solid rocket boosters and the capacity for placing 130 tonnes in Earth orbit or sending 46 tonnes (101,000lb) to the Moon. However, it would not be ready before *Artemis IX* in the early 2030s.

The SLS had been considerably delayed since it was first authorised in 2010, at which point its first flight was expected for December 2016. That date slipped progressively as a result of technical issues, funding problems and changes to mission requirements. Following an extensive sequence of manufacturing and assembly tests, the first core stage was taken from the Michoud Assembly Facility in New Orleans to the Stennis Space Center in Mississippi during January 2020 for what is called the Green Run. The stage was to have been test-fired for the duration it will be required to operate during a launch, about eight minutes in all.

The first attempt, on 16 January 2021, was aborted at 67 seconds due to conservative instrumentation specifically designed for the test. It worked as expected for the second attempt on 18 March, where the engines were gimballed and operated to higher levels than required in flight. The core stage left Stennis on 24 April and arrived at the Kennedy Space Center (KSC) three days later. Final assembly and checkout was completed on 12 June 2021.

Delivered in July 2017, the ICPS upper stage had been the first to arrive at KSC, where it was placed in storage. The SRB segments arrived by train on 12 June 2020, with the adapter element for mating

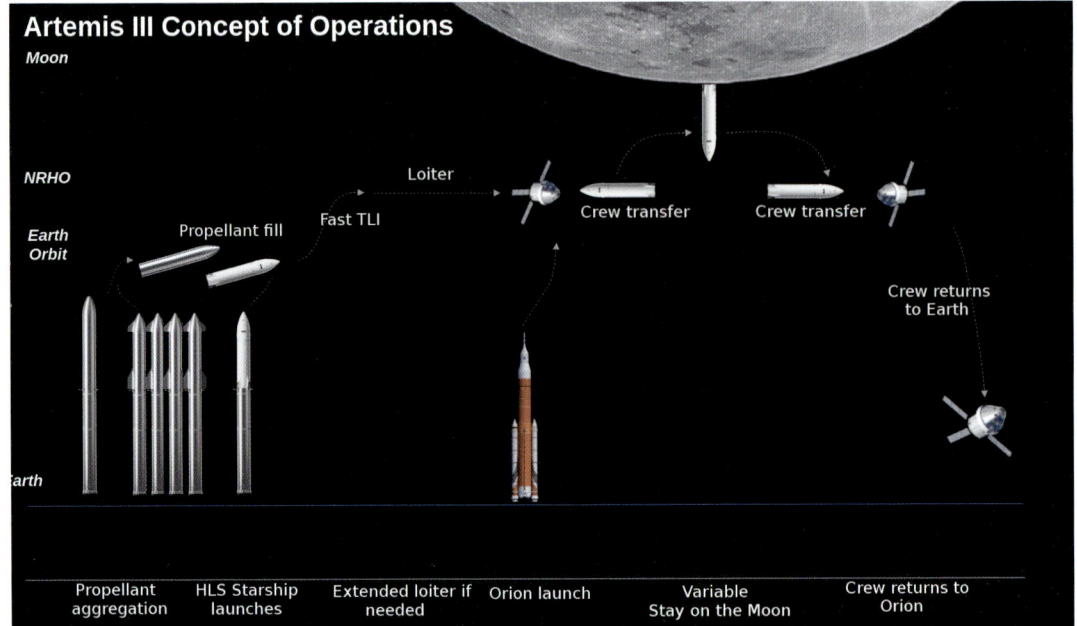

Above: By choosing *Starship*, NASA is committed to a complex procedure for each Moon landing, involving up to six launches by SpaceX and a single SLS/*Orion* flight to get up to four people on the lunar surface. (NASA)

Left: Currently, *Starship* is under development as the upper element to a heavy-lift launcher with more than twice the thrust of the *Saturn V*. (Starship)

the two stages on 29 July. Gathered in High Bay 3 of the capacious Vehicle Assembly Building (VAB), stacking of the booster segments began on 23 November, with the core stage placed between them and attached on 12 June 2021. Ten days later, the interstage adapter was in place, followed by the ICPS on 6 July. Connecting the SLS to the European Service Module, the *Orion* adapter was installed on 8 October 2021. *Orion* CM-002/ESM-01 was stacked 12 days later and preparations for roll-out began.

The *Orion* spacecraft had evolved from the initial design studies of the *Constellation* programme but remained essentially unchanged. As built for *Artemis*, it consisted of a Crew Module and the European Service Module with a Launch Abort System (LAS) on top for crew safety. The gross lift-off weight for the three combined elements stacked for *Artemis I* was approximately 32,296kg (71,200lb), of which the Crew Module weighed 9,300kg (20,500lb) and the ESM weighed 15,286kg (33,700lb). The LAS weighed 7,711kg (17,000lb).

To get into orbit, *Starship* is launched on top of a super-heavy booster designed to place large loads into space, as seen here. (SpaceX)

Consisting of a solid rocket motor delivering a thrust of 1,779kN (400,000lb) and a fairing completely covering the Crew Module, the LAS provides escape for the crew during ascent to orbit by separating the CM from the ESM and out-accelerating the giant SLS, should it run into trouble. It has more thrust than the *Atlas* rocket that launched the first American astronaut into orbit in 1962. The LAS is jettisoned by a less powerful solid rocket motor about 6 minutes 51 seconds after launch, when the ESM takes over for abort modes, carrying the *Orion* spacecraft into orbit should it need to escape from a potential explosion in the SLS.

With a base diameter of 5m (16½ft) and a height of 3.35m (11ft), the Crew Module has undergone several engineering improvements, including reforming the design of the structural pressure shell into seven separate pieces rather than the 13 welded sections it originally had, thus saving 317kg (700lb) in weight. Four crew seats are attached to the aluminium floor structure known as the 'backbone', a nine-piece bolted assembly criss-crossing the broad base of the spacecraft.

As finally designed after EFT-1, the base heat shield is made up of 186 blocks of Avcoat, with the sloping cone-shaped forward section supporting 1,300 black thermal protection tiles known as AETB-8, of the type used on the Shuttle Orbiter. After the Crew Module separates from the European Service Module at the end of the mission and prior to re-entry, attitude control is maintained by 12 rocket motors, each with a thrust of 711N (160lb).

Crucial to the success of *Orion*, its avionics system includes two management computers provided by Honeywell, each of which has two flight computer modules (FCMs) for flight control and software functions. Each FCM has internal redundancy for self-check, test and repair by disregarding the failed unit. These contingencies, which self-adjust the calculations after 22 seconds, accommodate potential outages during solar radiation storms in deep space and outside the Earth's magnetosphere.

Guidance and navigation equipment and software is at the core of commanding and controlling *Orion*'s attitude, orientation and propulsive manoeuvres to get to the next objective, and that includes GPS during low Earth orbit operations. NASA intends to remove the avionics equipment from CM-001 and install it in CM-002 for *Artemis II*. Absent from the *Artemis I* mission were the environmental control system, waste management equipment and crew suits. Standing in for humans and for measuring acceleration loads and radiation levels were three mannequins, one of which was named *Captain Moonikin Campos* after an *Apollo* flight engineer, and two instrumented torsos named *Zohar* and *Helga*.

The European Service Module (SM-1) has the same maximum diameter as the Crew Module and, with a height of 4.78m (15.7ft), it provides all the environmental control, electrical power, communications and propulsion for all phases of a typical mission. Electrical power is provided by four

Left: SpaceX personnel give scale to the size of *Starship* as the two stages are mated in a test at the Boca Chica launch site in Texas. (SpaceX)

Below: The core stage of the Block I Space Launch System (SLS) with its primary and secondary elements. (NASA)

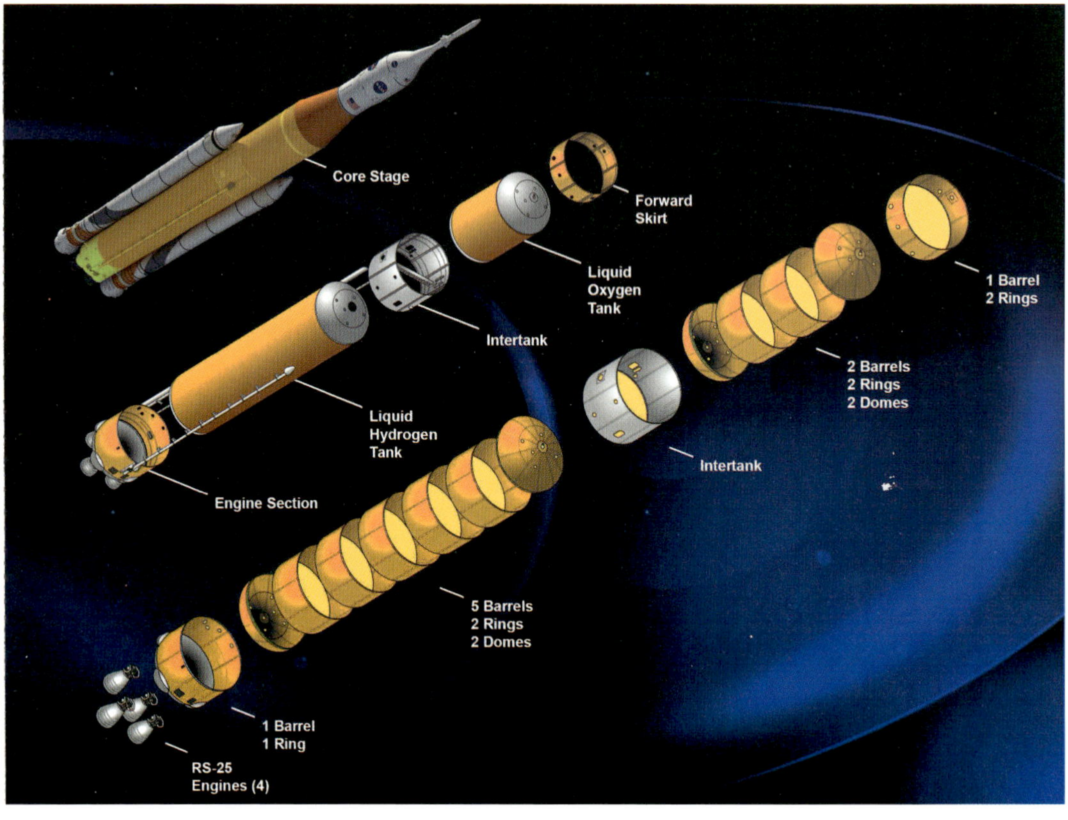

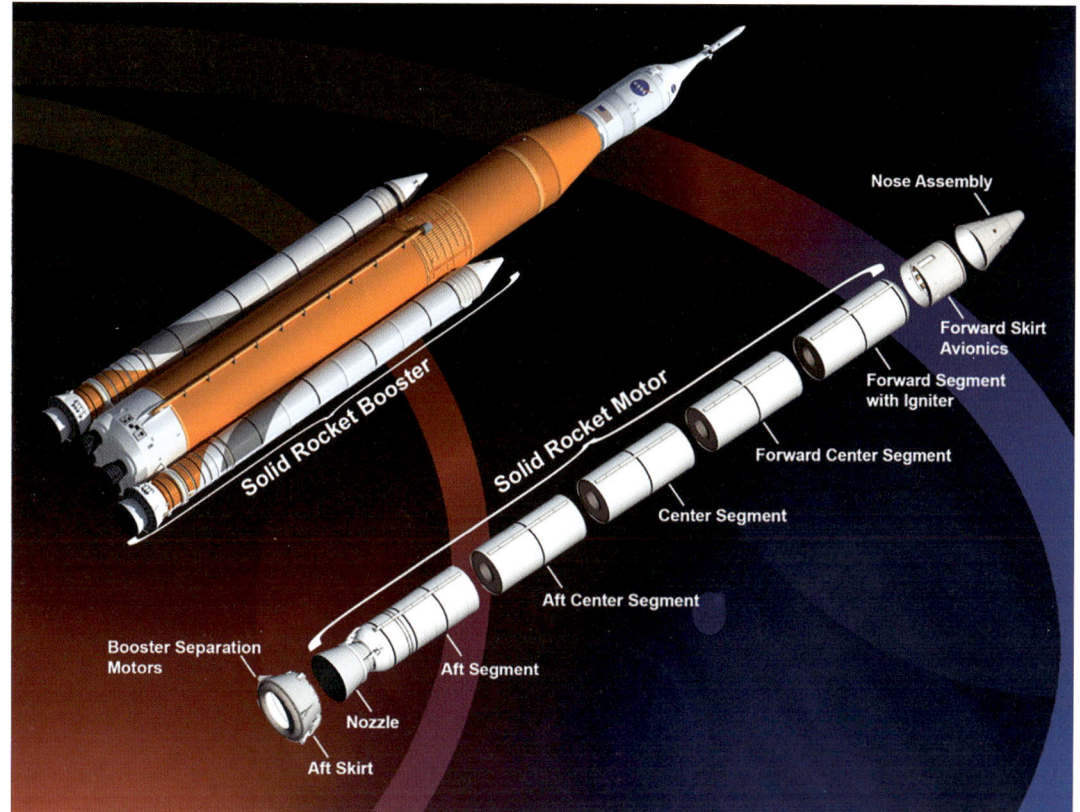

The five-segment Solid Rocket Boosters were a significant development on the SRBs used in the Shuttle, increasing thrust and extending burn time. (SLS)

solar panels, each 2m (6½ft) wide, and when deployed they have a length of nearly 7m (23ft). Arranged in a cruciform array, when deployed they span 18.9m (61ft) and support 15,000 solar cells providing a maximum 12kW of electrical power. All these services are provided for *Orion* until the ESM is jettisoned shortly before re-entry.

The propulsion system consists of a single rocket motor for main course corrections and orbital manoeuvres. With a thrust of 26.688kN (6,000lb), it is the motor used by the Shuttle for a similar function, where it was known as the Orbital Manoeuvring System (OMS); two of them were placed at the rear. The motor on SM-1 had flown on 19 Shuttle missions. An additional eight rocket motors, each with a thrust of 489N (110lb), provide auxiliary propulsion. Designated R-4D-11, these are derived from the R-4D motors used on the *Apollo* Service Module. A further 24 reaction-control thrusters each provide 222N (50lb) of thrust for attitude control.

The primary characteristics of the SLS have been described previously, but when stacked with *Orion* and its abort system, the assembly has a height of 98.3m (322.4ft) and a fuelled weight of 2.6 million kg (5.74 million lb). It has a lift-off thrust of 39,144kN (8.8 million lb) from its four RS-25 engines and the two SRBs. The first four SLS flights will use 16 RS-25 engines from the Shuttle programme, which collectively have already logged 25 missions. More engines are already under contract for later flights, beginning with *Artemis V*.

Each SLS Solid Rocket Booster has an additional fifth segment, which adds thrust and extends the burn time but here too, reusability from the Shuttle programme saves both manufacturing and money.

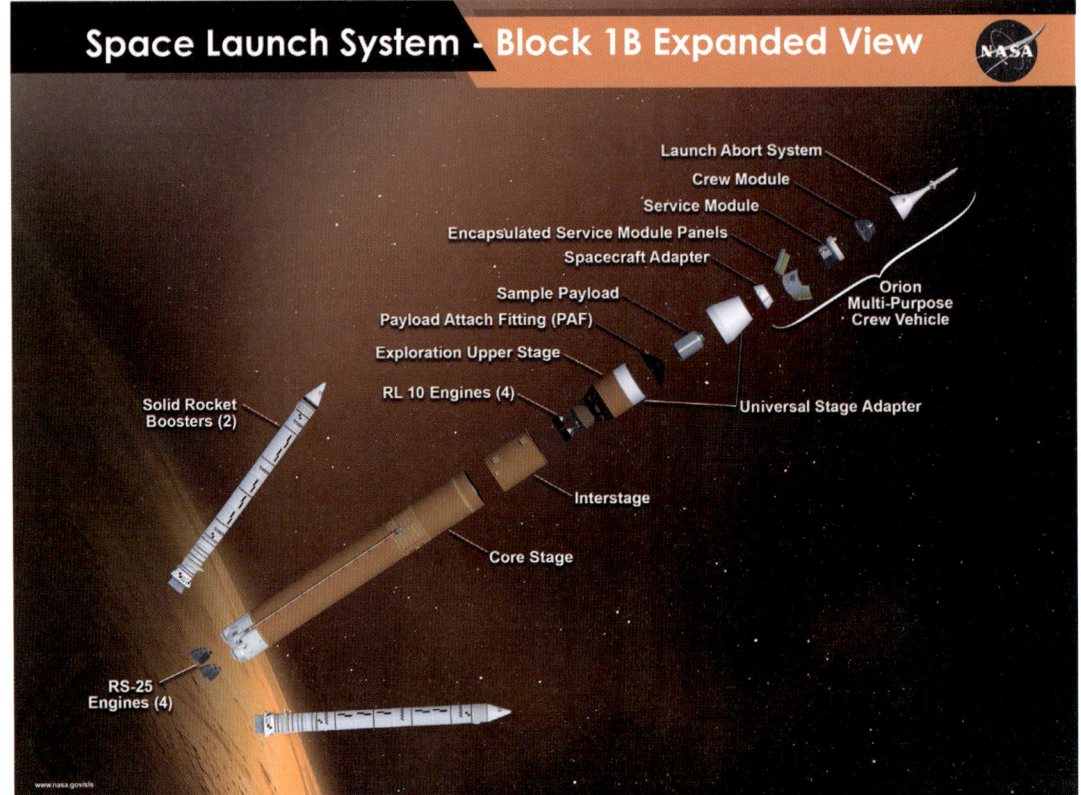

With its more powerful Exploration Upper Stage (EUS), the SLS Block IB will significantly improve the payload capability of the launch vehicle. (NASA)

In all, various elements of the two *Artemis I* boosters have flown a total of 97 Shuttle missions but on SLS flights the boosters will not be recovered. Contracts exist for additional SRB components and segments for later flights.

Artemis I Flies

Preparation for the first unmanned flight of SLS/*Orion* began when the assembled stack was rolled out from the Vehicle Assembly Building (VAB) to Launch Complex 39B (LC-39B) on 17 March 2022, the same launch pad where four *Apollo* missions and 53 Shuttle flights began. To test all the mechanical access arms, fluid transfer lines, electrical connections and propellant-loading equipment, a wet dress rehearsal (WDR) was planned for 3 April. This would involve loading all the propellants in a sequence typical of a live launch.

Things began to go wrong when a pressurisation programme stopped operations, and a second attempt on the following day was also aborted. A third attempt was similarly scrubbed when an helium pressurisation valve on the SLS upper stage failed. Replacing that component required the stack to be returned to the VAB on 26 April. It was not back at LC-39B until 6 June, after several checks and modifications had been made to some other components. There was concern over these delays, because there was a limit to the amount of time the SRB segments could remain stacked before launch and several waivers had to be applied as the overly conservative limit came and went.

Finally, on 20 June the WDR was completed to T-29 seconds, still short of the T-9.3sec at which point ignition of the three RS-25s in the core stage would have begun the launch sequence. A further

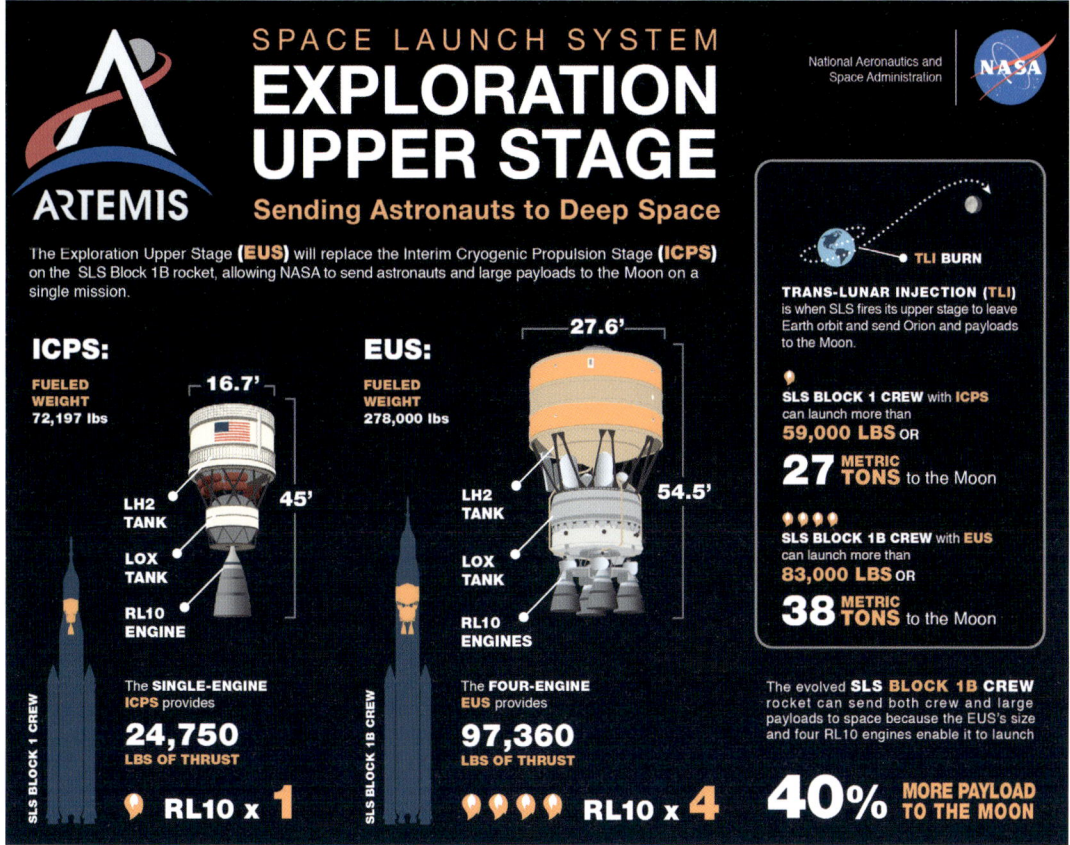

The key to the Block IB is the Exploration Upper Stage, with a 40 per cent increase on the payload capability of the Block I SLS. (NASA)

issue involved a hydrogen leak, which could be fixed when the rocket was rolled back into the VAB a second time on 2 July, this time for final pre-launch preparations.

The stack was back at LC-39B on 17 August, and five days later the critical Flight Readiness Review was held to clear the vehicle for flight. But it would be three months before the launch was successfully under way as further technical issues, lightning strikes on the pad and a hurricane warning got in the way. All this was in addition to restrictions on the launch window driven by the Earth-Moon geometry with respect to the Sun, plus other critical issues driven by flight requirements.

Artemis I was designed to stress the spacecraft, the equipment and the launch and flight operations' teams far beyond what they would be prepared to risk on a manned flight. Engineers wanted to know how well the systems behaved under, and sometimes above, design limits. The only safe way to test it to its maximum potential was without humans on board. Much of what would be learned from this flight would help write the rule book on how to fly *Orion* and what to do if things went wrong.

Ignition of the four RS-25 engines in the core stage occurred 6.4 seconds before the two SRBs. Lift-off occurred at 01.47.44 hours EST, 16 November 2022. The night flight of *Artemis I* created a spectacular display of fire, light and bone-shaking thunder as it soared ever higher into the sky. No rocket more powerful had ever lifted off from the surface of the Earth and made it all the way into space.

Little more than one minute later, it went through what engineers call Max-Q – the point of maximum aerodynamic pressure on the vehicle and the moment of greatest stress on the ascending stack. As it climbed, the atmosphere became thinner, reducing aerodynamic forces. Shortly after 2 minutes elapsed time, the SRBs separated at a speed of 5,100kph (3,170mph) and a height of 48km (29.9 miles). The remaining elements continued on their way as the protective shrouds encapsulating the aft end of the ESM were jettisoned a minute later and the Launch Abort System separated at 3 minutes 30 seconds. About 8 minutes 20 seconds after lift-off, the core stage shut down, its propellant expended, and separated 12 seconds later at a height of 167km (104 miles).

Next, the four solar arrays on the ESM were deployed in a procedure lasting about 12 minutes, Mission Control quickly reporting electrical power flowing from that system. To prepare for the Trans-Lunar Injection (TLI) burn that would send *Orion* towards the Moon, about 53 minutes after lift-off the ICPS upper stage fired for little more than 20 seconds to more closely circularise the orbit from its initial, elliptical path, this being known as a perigee raise manoeuvre. The critical 18-minute TLI burn was performed by the single J-2 on the upper stage, beginning 1 hour 26 minutes after lift-off over the continental USA. *Orion* separated from the ICPS 1 hour 54 minutes after launch.

The main engine on the ESM was fired for the first time at 7 hours 45 minutes, to test its operation and conduct the first trajectory correction burn. A second correction burn used the smaller thrusters to check those out at 06.32hrs EST on 17 November, 28 hours 45 minutes after launch. A third correction burn was completed at 07.12hrs EST on 20 November, when the auxiliary thrusters fired for six seconds, changing velocity by 3.7kph (2.3mph).

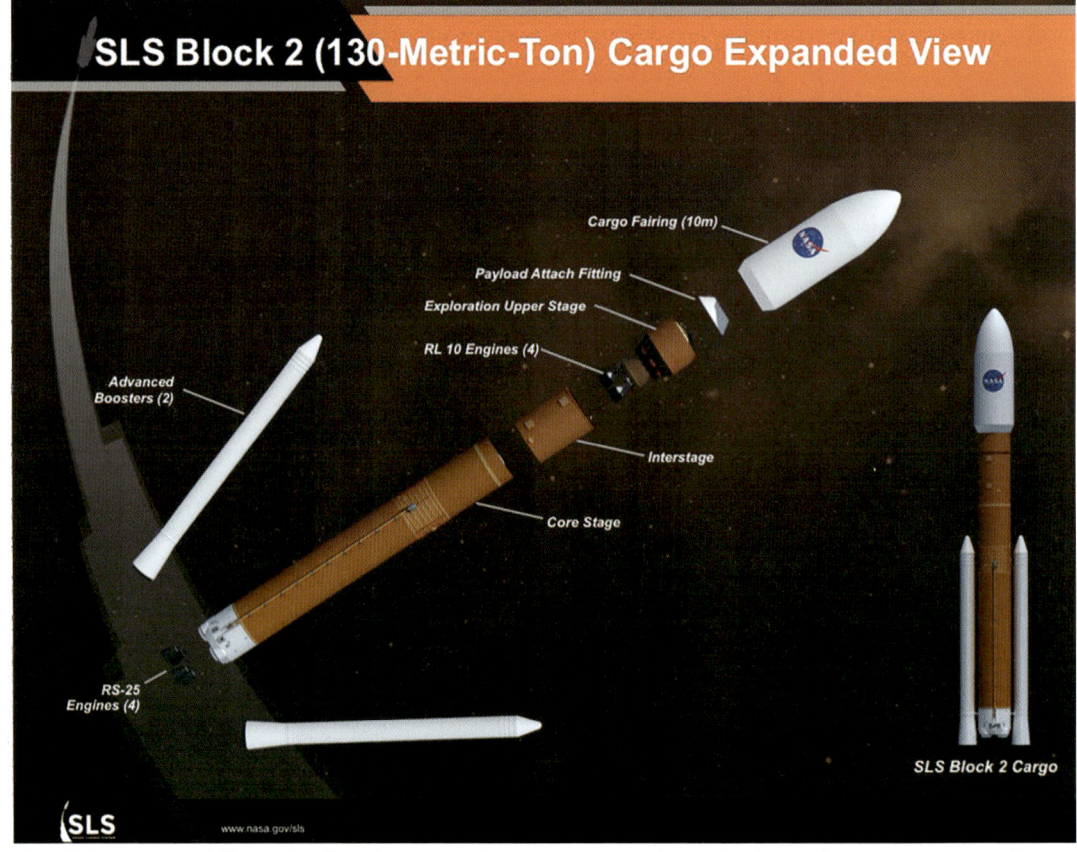

The Block 2 SLS would have significantly improved SRBs, increasing the payload capacity 71 per cent over the basic Block I. (NASA)

A Solid Rocket Booster is prepared for a test in July 2022 to improve its performance at Northrop Grumman's Promontory, Utah, facility where night workers give scale to the 727 tonne (1,600,000lb) rocket. (Northrop Grumman)

The 2022 development test of the five-segment Solid Rocket Booster unleashes a thrust of 16,13kN (3,600,000lb). (NASA)

All the way out to the Moon, the Earth's gravity had been pulling on *Orion*, inexorably slowing it from its initial speed of 36,300kph (22,500mph) after the TLI burn. At 14.09hrs EST on 20 November *Orion* passed into the Moon's gravitational sphere of influence, at which point it was travelling at little more than 644kph (400mph) and would start speeding up. A fourth correction burn was conducted at 02.44hrs EST on 21 November, using the auxiliary thrusters to fine-tune the flight path.

Throughout the long transit, engineers continued to perform a wide range of tests to monitor the way *Orion*'s systems responded to a variety of different operating modes, some of which would never be knowingly used on a manned flight. There were a handful of issues, but none were critical and in total they were fewer than expected.

As the spacecraft neared the Moon, flight controllers geared themselves for the critical burn that would secure *Orion* in lunar orbit, the Moon's pull increasing speed to a maximum 8,082kph (5,032mph) at the time the ESM engine fired for *Orion* to enter an elliptical orbit. In Mission Control there was undivided attention on the critical upcoming burn. But that would happen on the far side of

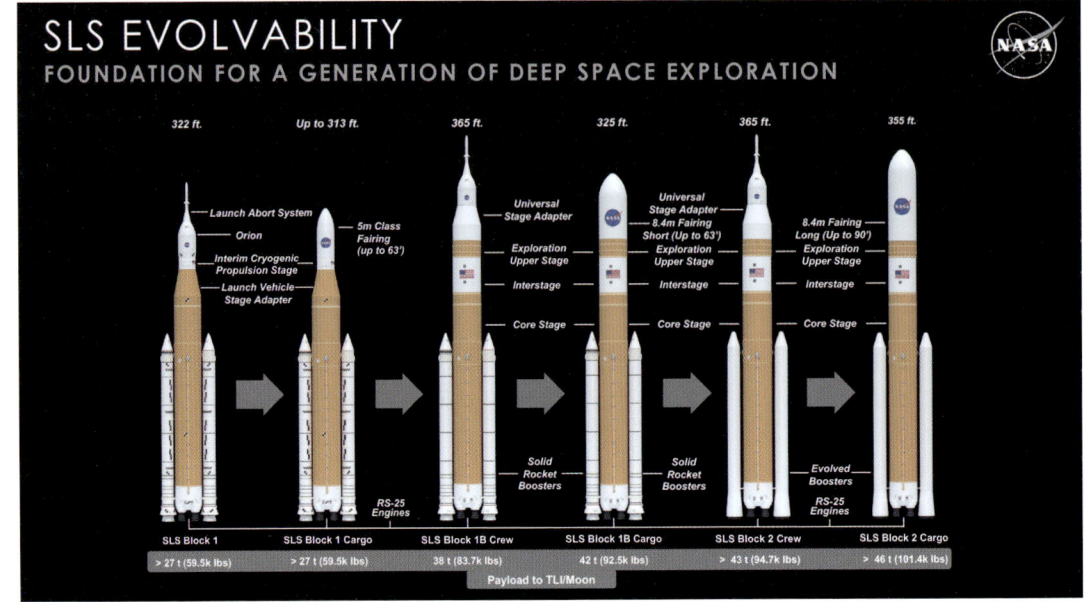

Above: The capacity for the SLS to evolve had been implicit within the design and is represented by the planned increase in performance. (NASA)

Left: Preparation for the first SLS launch with *Artemis I* included acoustic tests with the *Orion* spacecraft and the European Service Module, here displaying its folded solar arrays. (NASA)

the Moon, out of radio contact with Earth. Ground controllers lost contact as *Orion* passed around the left-hand side of the Moon as viewed from Earth at 07.25hrs EST on 21 November, a retrograde path because it was travelling in the opposite direction to the orbit of the Moon around the Earth.

At loss of signal, *Orion* was 1,500km (900 miles) from the Moon but, 19 minutes later, the main engine fired for 2 minutes 30 seconds at an altitude of 527km (328 miles), slowing the spacecraft by 644km/hr (400mph). This Outbound Powered Flyby manoeuvre placed *Orion* only 130km (81 miles) above the surface, quickly climbing up again in an elliptical path from where it would be placed in its Distant Retrograde Orbit (DRO).

Signal was acquired when *Orion* appeared around the right-hand limb of the Moon at 07.59hrs EST. As it climbed ever higher, at 08.37hrs EST *Orion* passed over the *Apollo 11* site at a height of 2,252km (1,400 miles) and then over the Apollo 14 site at 9,654km (7,000 miles) before crossing the Apollo 12 site 12,389km (7,700 miles) below. The following day, 22 November, at 00.01hrs EST, the fifth trajectory correction burn was accomplished by firing the auxiliary thrusters for 5.9 seconds, changing speed by

The solar arrays on the European Service Module are extended in a test of the deployment equipment. (ESA)

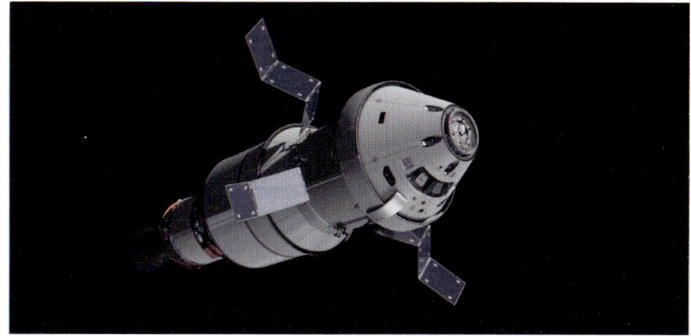

Folded against the cylindrical side panels of the European Service Module for launch, the solar arrays are deployed in space. (ESA)

3.5km/hr (2.18mph). At 10.49hrs EST, *Orion* exited the Moon's gravitational sphere of influence at a distance of 64,349km (39,993 miles) from the surface.

The sixth correction burn occurred at 16:52hrs EST on 24 November, lasting 17 seconds and changing speed by 9.7km/hr (6mph). All the while, continuous engineering tests and checks were made of several systems, including the star trackers, devices that utilise the different brightness levels of selected stars to give *Orion* information about its orientation. Cameras attached to the tips of the four solar arrays were also used to observe the condition of the thermal insulation tiles on the Crew Module and to look for any tiny impacts on the ESM as well as capture views of the Moon.

Orion at a distant place

More than four days after achieving lunar orbit, a further burn of the EMS main engine was made at 17.52hrs EST on 25 November, lasting 88 seconds for a velocity change of 398km/hr (247mph). This raised the low point of the orbit (perigee) and inserted *Orion* into the stabilised and circular DRO path, achieving maximum distance from the Moon of 64,000km (40,000 miles) at 16.05hrs on 28 November. It would take the spacecraft almost a week to complete half an orbit of the Moon, from which point it would begin the sequence to come home but, prior to that, the records started tumbling.

At 08.42hrs on 26 November, as it climbed ever higher in its elliptical path, *Orion* surpassed the previous record for the distance from Earth any spacecraft built to carry humans had travelled, although on this flight it was without a crew on board. The now broken record had been set by *Apollo 13* in April 1970, when it reached 400,171km (248,655 miles) from Earth on its slingshot flight around the Moon.

Propellant tanks for the ESM-2, which will support the *Artemis II* mission carrying astronauts around the Moon, probably in 2025. (ESA)

The main engine on the European Service Module is from one of the Shuttle Orbiters and the eight auxiliary rocket motors here, with red caps on, are a development of the type used on *Apollo*. (ESA)

Later, on 26 November, an 'orbit maintenance' burn took place at 16.52hrs EST when *Orion*'s auxiliary thrusters fired for less than a second to adjust the path and change speed by 0.5km/hr (0.32mph). Another record was achieved shortly after 15.00hrs EST on 28 November, when at 432,210km (268,563 miles), *Orion* was at its greatest distance from Earth. A second orbit maintenance manoeuvre occurred on 30 November when the smaller thrusters were fired for 95 seconds, longer than really necessary but extended so that engineers could test a longer burn than the maximum previously achieved of less than 17 seconds.

The sustained operation of these highly precise velocity and course adjustments were not to correct previously flawed manoeuvres, but to compensate for minor perturbations caused by irregularities in the Moon's mass. Like Earth, the Moon is not perfectly uniform and has slightly different gravitational anomalies from place to place, which pull and tug at orbiting spacecraft. Learning how to navigate these was one of the objectives of this mission.

The exit from the DRO path began at 16.53 hours EST on 1 December with the first of two critical burns. The main engine on the ESM fired for 1 minute 45 seconds, changing speed by 498km/hr (309.5mph) and pulling the orbit back into the elliptical path it had departed six days before. A trajectory correction burn was conducted at 22.54hrs EST for five seconds, changing the velocity by

The SLS Interstage separating the two propellant tanks in the core stage, during manufacture at the Michoud Assembly Facility. (NASA)

The core stage for *Artemis I*, completed in November 2019. (NASA)

Roll-out of the *Artemis I* core stage, given scale by the ground-handling crew as it leaves for the Kennedy Space Center. (NASA)

Above left: Preparation of *Artemis I* begins with Solid Rocket Booster segments being stacked in the Vehicle Assembly Building. (NASA)

Above right: Both Solid Rocket Boosters for *Artemis I* are assembled, ready to receive the core stage. (NASA)

Left: The massive core stage of *Artemis I* arrives inside the Vehicle Assembly Building. (NASA)

0.48km/hr (0.3mph). *Orion* re-entered the lunar sphere of influence at 17.45hrs EST on 3 December when 64,349km (39,993 miles) from the surface. It performed the second course correction on this leg at 11.43hrs EST on 4 December, using the auxiliary thrusters to increase velocity by 1.86km/hr (1.16mph).

On 5 December, a few hours before the burn that would bring *Orion* home, a minor correction at 05.43hrs EST involved the small attitude control thrusters firing for 20.1sec, changing velocity by

Above left: Elevated to a vertical position, the core stage is lowered between the Solid Rocket Boosters to which it will be attached. (NASA)

Above right: The adapter is attached to the top of the core stage ready for the upper stage. (NASA)

Below left: The upper stage for SLS is lowered into the adapter, to the top of which will be attached the *Orion* and its European Service Module. (NASA)

Below right: *Artemis I* rolls out to the LC-39B launch pad, a journey it would repeat before lift-off when technical problems needed attention. (NASA)

2.24km/hr (1.39mph) to determine their potential, should the larger thruster rockets fail. The second and final close pass across the lunar surface occurred at 11.43hrs EST and an altitude of 130km (80.6 miles). Known as the Return Powered Flyby, the main engine on the ESM fired up for 3 minutes 27 seconds to change speed by 1,054km/hr (655mph). This placed it on a path swinging around the Moon one last time for the journey back to Earth.

Less than a day after departing lunar orbit, *Orion* left the Moon's sphere of influence for the last time at 02.29hrs EST on 6 December. A return trajectory correction burn was completed at 05.43hrs EST where the small attitude thrusters fired for 5.7 seconds, changing the speed of *Orion* by 0.6km/hr (0.4mph). Again, cameras were used to map the exterior surface of the Crew Module for signs of damage to the thermal tiles and to generally survey the exterior for possible micrometeoroid impacts.

Left: *Artemis* at LC-39B, a site modified to support the Space Launch System for the next evolution in deep-space exploration. (NASA)

Below left: Lift-off! A night launch in the early hours of 16 November 2022 as *Orion* begins a four-week journey to the Moon and back. (NASA)

Below right: Launch director Charlie Blackwell-Thompson in the Firing Room at the Launch Control Center alongside the Vehicle Assembly Building. (NASA)

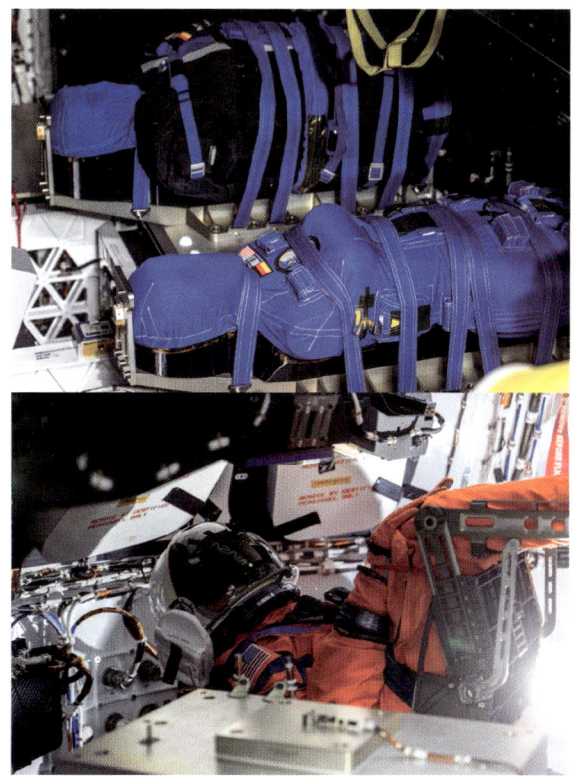

Right: The Moonikin mannequin (bottom) and the instrumented torsos Helga and Zohar, placed in *Orion* to collect in-flight data. (NASA)

Below: Moonikin in a crew couch of the kind astronauts will use on *Artemis II* and subsequent flights. (NASA)

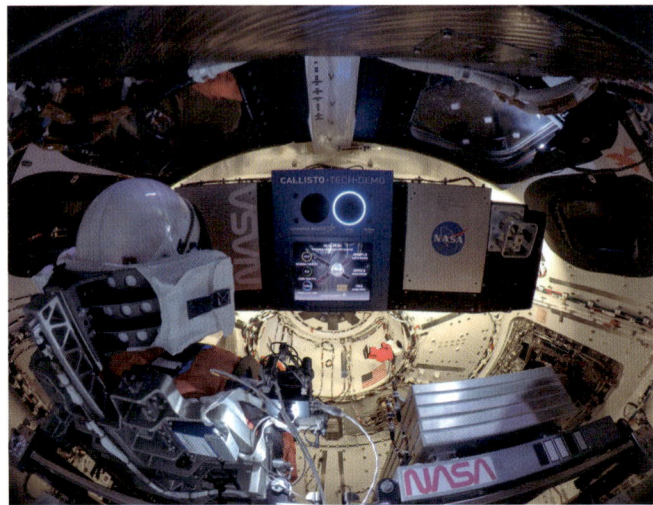

Left: A shot taken inside *Orion* on Day 11 after entering lunar orbit, with displays as they would be on a manned flight. (NASA)

Below: The flight trajectory and mission map for *Artemis I*. (NASA)

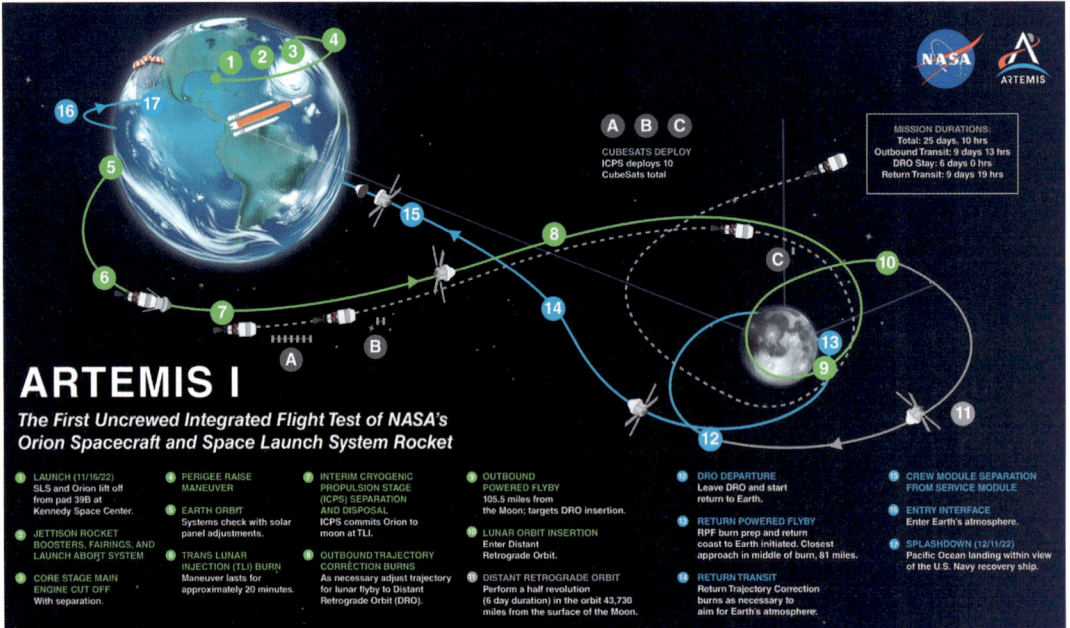

Other tests included measuring the amount of 'sloshing' that occurs in the propellant tanks under microgravity conditions when the spacecraft is moved around.

Engineers also conducted tests on the thrusters under different conditions and with varying quantities of propellants in the separate tanks. Frequent monitoring of use versus quantities remaining found that propellant management had been good. The ESM is designed to carry 8,600kg 18,960lb) of usable nitrogen and monomethyl hydrazine, each propellant in two 2,000-litre (528 US gallon) tanks.

As a typical set of measured quantities on the way back and as an example among several parameters, engineers noted that 5,470kg (12,060lb) had been used, 97.5kg (215lb) less than expected, with 991kg (2,185lb) reserved as a margin, 124kg (275lb) more than predicted before flight. These values were

Right: A view of Earth from one of the cameras on the exterior of *Orion*, showing the main propulsion system and some of the auxiliary thrusters on the European Service Module. (NASA)

Below left: The upper section of the European Service Module and the *Orion* Crew Module, showing the ingress/egress hatch and some of the attitude control thrusters. (NASA)

Below right: Another view of *Orion*, showing the opposite side to the entry hatch with the umbilical shroud over conduits and plumbing connecting the two modules. (NASA)

continuously monitored as the mission moved towards completion, as were all the consumables on board. Other consumables normally carried in the ESM at launch would be a total 240kg (529lb) of water in four tanks, 90kg (198lb) of oxygen in three tanks and 30kg (66lb) of nitrogen in one tank.

On 9 December, the attitude control thrusters on the Crew Module were activated and pulse-fired for 75 milliseconds in opposing pairs, to ensure what engineers call a non-propulsive vent in that no change in speed was made to disrupt the trajectory. The effective operation of thrusters on the CM was essential to ensure that the spacecraft's guidance system kept it correctly aligned for the planned descent profile.

As *Orion* neared Earth, a further trajectory correction burn was conducted at 15.32hrs EST on 10 December with the auxiliary thrusters firing for eight seconds, speeding up the spacecraft by

Above left: A close-up of the aft end plate on the European Service Module, with the main rocket motor and auxiliary thrusters. (NASA)

Above right: A detailed look at the pivot and articulation connectors for the four solar panels. (NASA)

Left: As *Orion* comes around from the far side of the Moon, Earth is seen in the distant horizon. (NASA)

5.4km/hr (3.4mph). The final trajectory correction burn lasted eight seconds at 07.20hrs EST the following day to add 1.94km/hr (0.68mph) to *Orion*, now increasing speed due to the gravitational pull of Earth. Separation of the European Service Module took place at 12.00hrs EST and the two elements of *Orion* drifted apart to ensure a clear separation at re-entry.

The Crew Module sliced into the atmosphere at 12.20.14hrs EST, the planned splashdown spot having been 550km (342 miles) south to avoid bad weather. Notionally considered the beginning of the atmosphere at an altitude of 121.89km (75.75 miles), when the deceleration on the spacecraft measures 0.05g, the descent profile involved a skip technique. This minimises thermal stress on the heat shield and limits deceleration to 4g; *Apollo* re-entry profiles from lunar distance imposed up to 7.5g on the crew. No US spacecraft had flown this type of re-entry before, although the Russians had done so with their returning *Zond 7* spacecraft in August 1969.

Two periods of 'black-out' followed when plasma enveloped the Crew Module, blocking electronic communication before the spacecraft was down to 12,192m (40,000ft) and a speed of 523km/hr (325mph), 15 minutes 14 seconds after Entry Interface (EI). At EI+15 minutes 48 seconds the apex cover, weighing 454kg (1,000lb) and measuring 2.9m (9.5ft) in diameter, was jettisoned, immediately followed by the deployment of two drogue parachutes. At an altitude of 2,895m (9,500ft), and a speed of 209km/hr (130mph), three pilot 'chutes deployed, followed by the three main 'chutes at EI+17 minutes 12 seconds. Reefed initially, they were fully deployed as the Crew Module descended to a

Above: Only 9,650km (6,000 miles) from the Moon, the navigation camera gets a view of the surface. (NASA)

Right: Climbing ever higher on the far side of the Moon, *Orion* sees both Moon and Earth as spheres in the sky. (NASA)

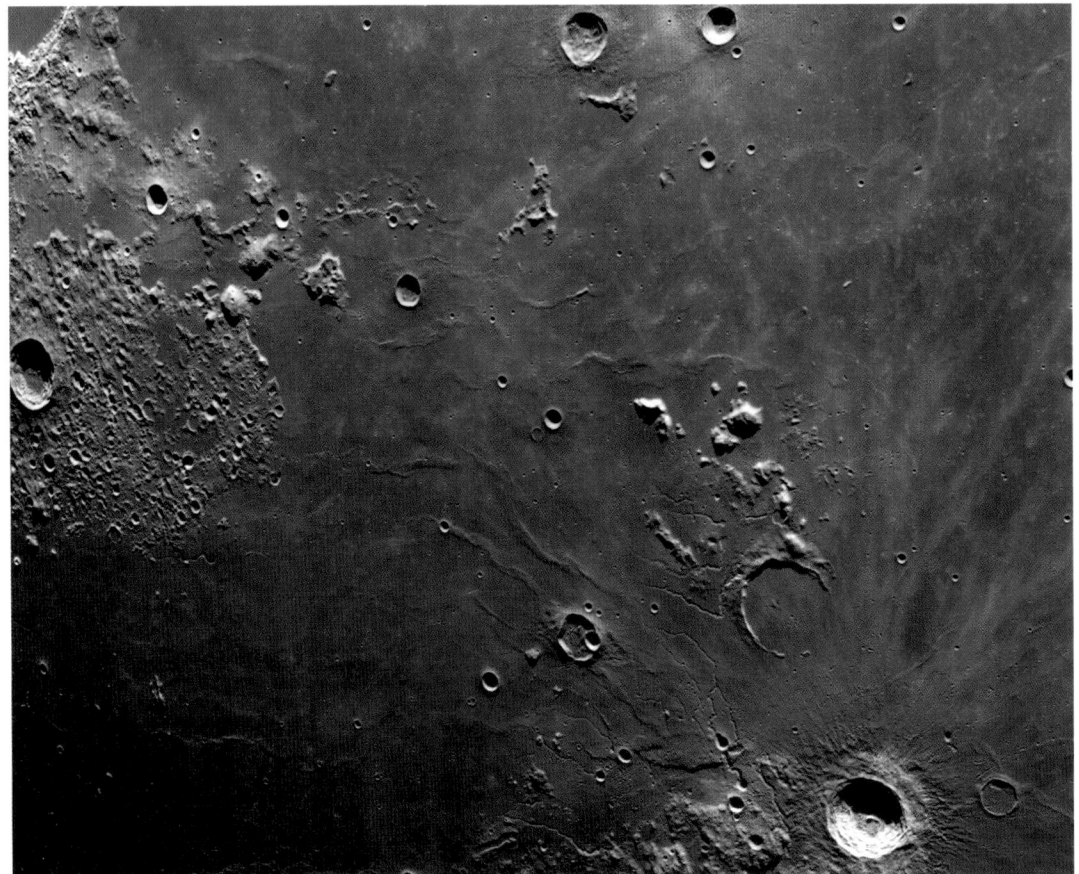
The second and final close pass across the surface of the Moon before the return journey to Earth. (NASA)

perfect splashdown almost three minutes later, at 12.40.30hrs EST and a speed of little more than 32km/hr (20mph). The mission had lasted 25 days 10 hours 52 minutes 46 seconds.

Next up

Orion reached the surface of the Pacific Ocean off Baja just 3.9km (2.4 miles) off target but within sight of the recovery ship *USS Portland*, well inside the 10km (6.2 miles) diameter circle in which it was expected to splash down. Before flight, 124 objectives had been set and many more were added as the flight progressed. Evaluation of all the data and analysis of the mission would continue well into 2023. But, new too, was the recovery technique, which is very different for *Orion* compared to *Apollo*.

To get the Crew Module back, the *USS Portland* provides a unique capability for containing the spacecraft without having to have it recovered by crane. An aft well-deck floods to allow the module to be winched in while still afloat, then is drained to facilitate access. *Orion* was left afloat for two hours before recovery, to collect additional data on soak-back heat levels through the structure before it was towed into the well-deck. *Portland* docked at San Diego on 13 December, where the spacecraft was offloaded for trucking to the Kennedy Space Center by the end of December.

Preparations for *Artemis II* and the first manned flight around the Moon in more than 50 years are well under way, with equipment for both the SLS rocket and the *Orion* spacecraft already at the Kennedy Space Center. NASA says it wants to send that crew around the Moon in 2024, but that is

Above left: A camera on one of the solar panels gets a view of the lit Crew Module interior. (NASA)

Above right: The return-to-Earth burn complete, *Orion* looks back at the Moon. (NASA)

unlikely. An independent government oversight report claims that it requires 27 months from the return of *Artemis I* to get all the avionics and other items moved across to CM-003, which will fly that mission. It is highly unlikely that *Artemis II* will launch before 2025.

The initial Moon landing by *Artemis III* is badged by NASA as a capability demonstration for getting people back on the Moon. Later landings will probably involve a more advanced version of the adapted *Starship* upper stage or be conducted by the alternative and more sustainable lander vehicles sought by the agency as a hedge against problems with the SpaceX lander. The schedule is nonetheless challenging and requires a lot to happen if a landing is to be achieved within the next few years.

Home to every US manned space flight since 1965, Mission Control at the NASA Johnson Space Center monitors the *Artemis I* flight. (NASA)

Above left: Splashdown in the Pacific Ocean on 11 December, more than 25 days after launch. (NASA)

Above right: NASA flight controller Julie Reed, serving as a flight dynamics officer (FIDO) in Mission Control. (NASA)

Below: With five inflatable bags, which would right the Crew Module should it be pulled apex-down by the parachutes, *Orion* awaits recovery. (NASA)

Orion is manoeuvred into the well deck on the USS *Portland*. (NASA)

The *Artemis II* mission will carry four astronauts around the Moon and back, possibly in early 2025. (NASA)

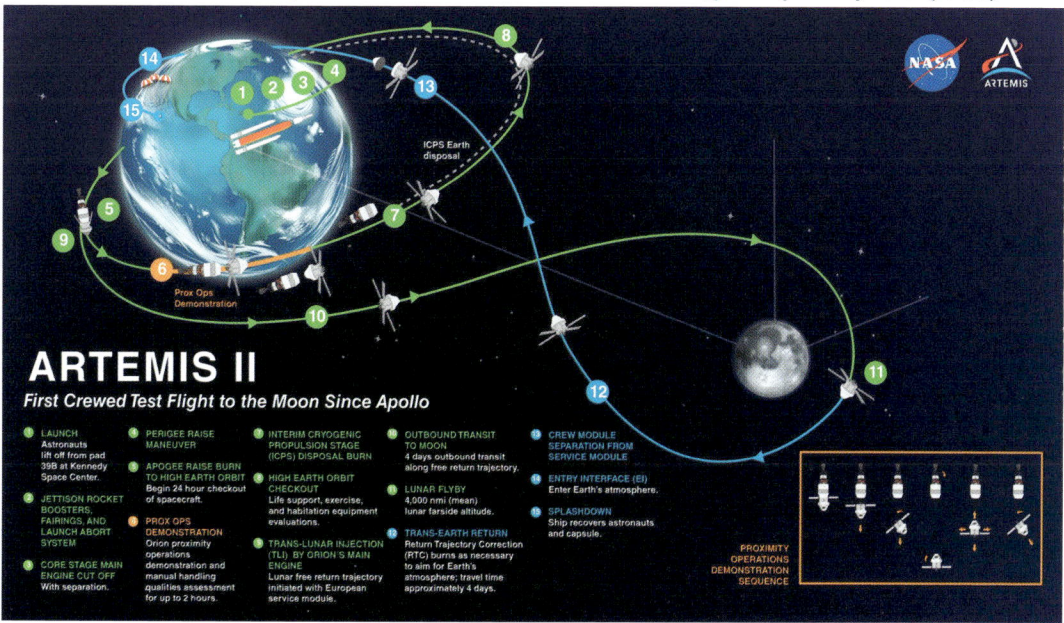

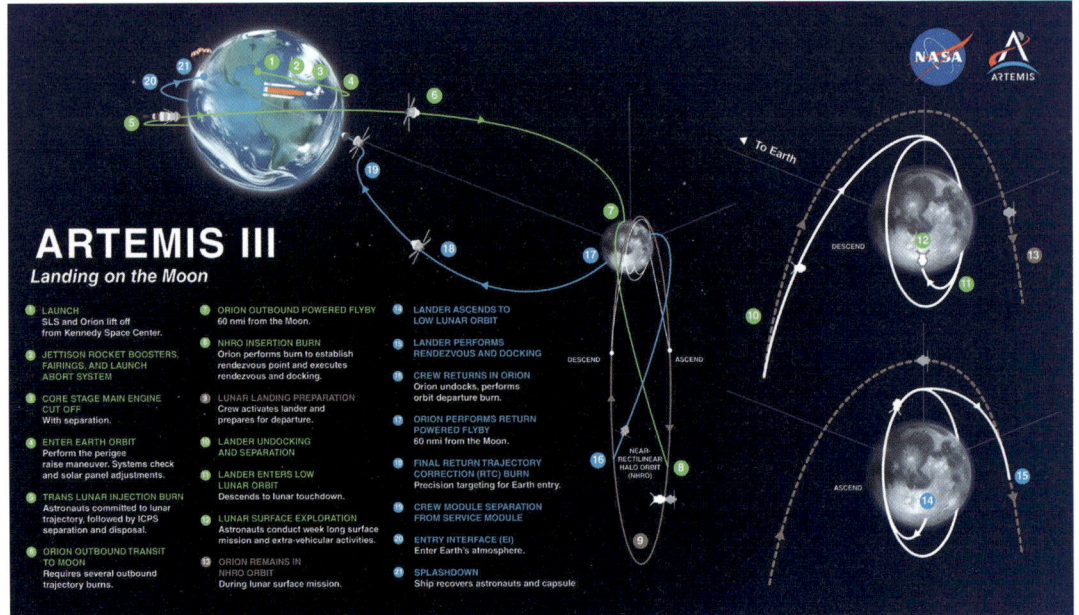

Artemis III is scheduled to put the first humans on the lunar surface since the last *Apollo* astronauts returned to Earth in December 1972, 51 years before the flight of *Artemis I*. (NASA)

Future missions to the Moon will use the Lunar Gateway as a staging post to the surface. Europe, Japan and Canada will provide elements of the station. (NASA)

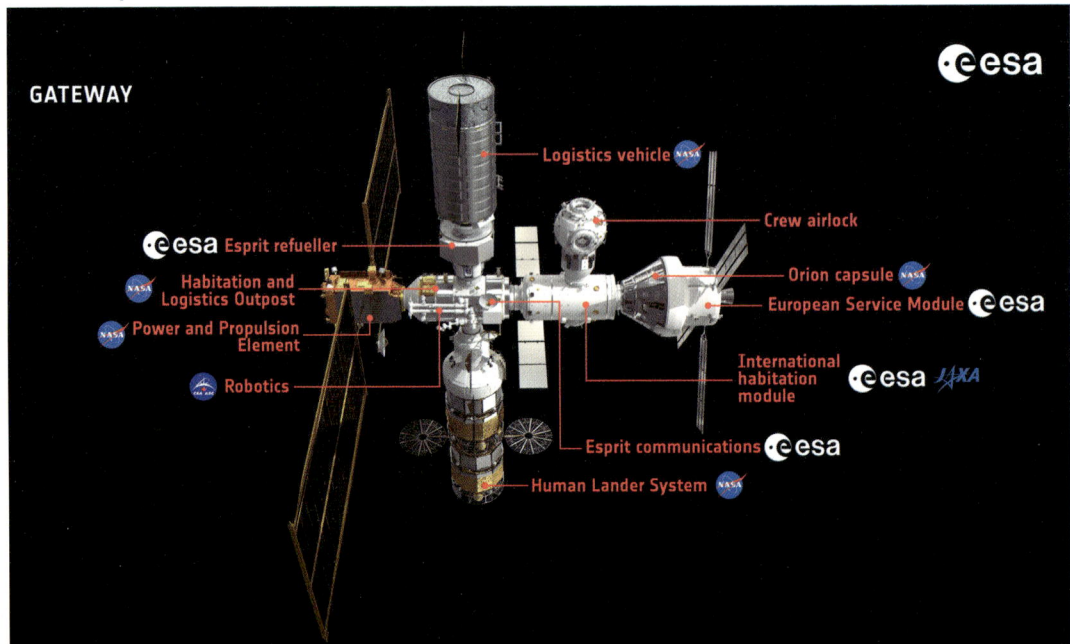

When SpaceX received the development contract for the Human Landing System, it was expected to conduct the first orbital launch of the massive, two-stage system from which the HLS would be developed in the first quarter of 2022. It was to be followed by an in-orbit propellant transfer test

The future for human space flight will thrive on shared investment with commercial companies and international cooperation between governments. (SpaceX)

during the third quarter of 2022, preceding a long-duration flight test in the first quarter of 2023. The first unmanned test landing on the Moon was scheduled for the fourth quarter of 2023, to qualify the concept.

By the beginning of 2023, the basic *Starship* had yet to achieve orbital flight and development was far behind schedule. Conversely, the heavy-lift Space Launch System and the *Orion* spacecraft, with its Lockheed Martin Crew Module and the European Service Module, were more or less on time for their own respective roles. Even given fair winds and following seas, it is still unlikely that the landing will occur before 2026.

Other books you might like:

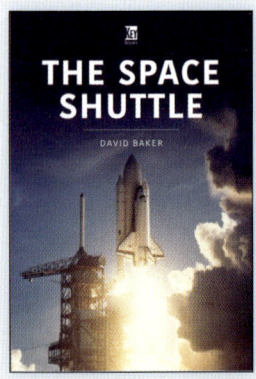
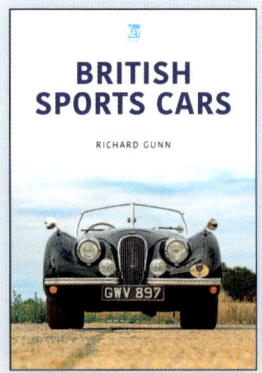

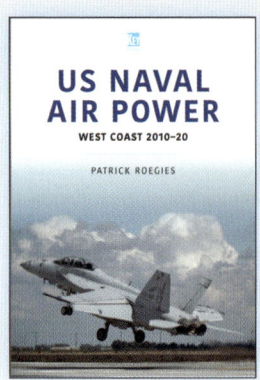

Air Forces Series, Vol. 2

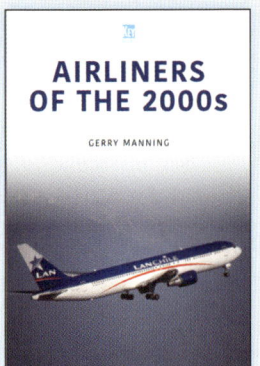

Historic Commercial Aircraft Series, Vol. 5

For our full range of titles please visit:
shop.keypublishing.com/books

VIP Book Club

Sign up today and receive
TWO FREE E-BOOKS

Be the first to find out about our forthcoming book releases and receive exclusive offers.

Register now at **keypublishing.com/vip-book-club**

Our VIP Book Club is a 100% spam-free zone, and we will never share your email with anyone else. You can read our full privacy policy at: privacy.keypublishing.com